Sexual Selection and Animal Genitalia

Sexual Selection
and
Animal Genitalia

William G. Eberhard

Harvard University Press
Cambridge, Massachusetts, and London, England   1985

This book is printed on acid-free paper, and its binding
materials have been chosen for strength and durability.

Library of Congress Cataloging in Publication Data

Eberhard, William G.
  Sexual selection and animal genitalia.

  Bibliography: p.
  Includes index.
    1. Sexual selection in animals.
2. Generative organs, Male.
I. Title.
QL761.E24   1985   574.1'662   85-907
ISBN 0-674-80283-7 (alk. paper)

To the memory of Bob Silberglied

Acknowledgments

A book that attempts to cover as many fields of knowledge as this one does necessarily suffers from an incomplete and unbalanced treatment of the literature, and I apologize at the outset to those authors whose papers I should have cited but did not. I owe most of whatever success I have had in finding key references and information to the many people who read and commented on preliminary versions of the manuscript, supplied references and unpublished work, and cheerfully allowed their brains to be picked: Annette Aiello, John Alcock, Warren Atyeo, William Brown, Lawrence Blumer, George Byers, Debbie Clark, David Clark, Jonathan Coddington, Ted Cohn, Jim Connor, David Cowan, Donald Davis, Janis Dickerson, Malcolm Edmunds, George Eickwort, Howard Evans, Jack Franclemont, Oscar Francke, Richard Freitag, Larry Gilbert, Robert Gordon, Darryl Gwynne, William Hamilton, Larry Kirkendall, Otto Kraus, Herbert Levi, Jim Lloyd, Yael Lubin, Robert Matthews, Tom Moore, Charles Michener, Nancy Moran, Gary Neuchterlein, Brent Opell, Mark O'Brien, Norman Platnick, Stanley Rand, Douglas Robinson, John Sivinski, William Sheehan, Pat Smith, Randy Thornhill, Gary Stiles, Robb Voss, Jeff Waage, Tom Walker, H. Wallace, Peter Weygoldt, Don Windsor, Stephen Wing, and Harald Witte. An unpublished manuscript by R. D. Alexander containing the seminal idea that male and female genitalia evolve in evolutionarily unending competitive races was especially important, and I thank him for sharing his ideas with me.

Barry OConnor was extraordinarily generous with his time in answering ignorant questions and telling wonderful mite stories, and Leonora Gloyd, Henry Hespenheide, Stewart Peck, Bill Shear,

Julian Shepard, and Jon Yates were unstinting with valuable information and references. I am particularly grateful to Crista Deeleman-Reinhold, K. C. Emerson, Thomas Barr, Roger Price, and Rowland Shelley, who gave thoughtful and extensive replies to questions "out of the blue" from someone they did not know; obtaining such disinterested help makes one proud to be a scientist.

Thanks are also due Barbara and Mike Robinson for their generous help, which allowed me to put this manuscript onto my own word processor and thus make the final preparation more fun.

It is not a casual coincidence that my wife, Mary Jane West Eberhard, was working on the relationship between sexual selection and the evolution of animal communication while I read and wrote on this project. She has been my most valuable colleague, and many of her ideas are incorporated in the book.

Much of the library work was done at the University of Michigan while I was a visiting scientist at the Museum of Zoology, and I thank the staff there for numerous courtesies. I am also grateful to Jack Marquardt and his staff at the Smithsonian Library. Financial help came from the Vicerrectoría de Investigacion of the University of Costa Rica. I thank Beate Christy, Maria Luz Jimenez, Maria Morelo, Kathy Stringham, and especially Ninotchka Franco Smith for help in preparing the tables and early versions of the manuscript, and Peg Anderson for numerous careful and perceptive editorial suggestions.

Contents

Sexual Selection and Animal Genitalia

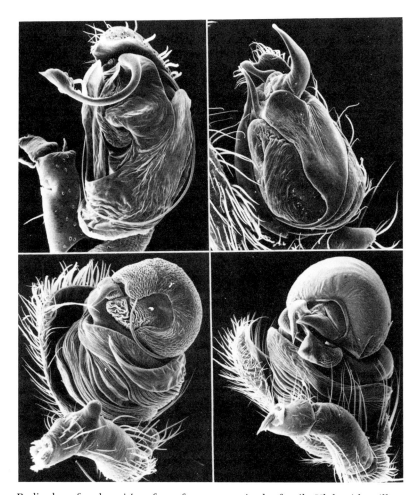

Pedipalps of male spiders from four genera in the family Uloboridae, illustrating the complexity and diversity of form typical of intromittent organs.
Male spiders inseminate females with these modified tips of leglike appendages, which are first primed with sperm from primary genitalic openings
on the male's abdomen. Pedipalps are key taxonomic characters in most
spider groups. (From Opell 1979.)

1 *An introduction to the problem*

It has long been recognized that animal genitalia and the structures associated with them are often very complex in form and that in a wide variety of animal groups, genitalia are especially useful in distinguishing closely related species. The aim of this book is to answer the question of why this is so. Since any structure that is a consistently useful taxonomic character at the species level must have evolved both rapidly and divergently (that is, it acquires a new form in each new species), the question can be rephrased as: why do animal genitalia so often evolve both rapidly and divergently?

It has not been generally appreciated that this question concerns one of the most basic tendencies in all of animal evolution. In this first chapter I will demonstrate the near universality of the pattern, surveying genitalic evolution in a variety of animal groups and illustrating the astonishing and apparently extravagant complexity of many genitalia. In Chapters 2–4 I discuss the hypotheses that have been proposed to explain genitalic evolution. In Chapters 6–11 I explain a new hypothesis, sexual selection by female choice, and discuss the evidence relating to it. The overall conclusion is that sexual selection by female choice accords better with the data than any of the other hypotheses.

Operational definitions

Genitalia. Before surveying different animal groups, it is necessary to deal with a troublesome semantic point. Male animals employ a remarkable assortment of structures in sperm transfer, and these structures are derived from a diverse array of organs. Some structures are located very close to the male gonopore, but others are not, so the distinction between genitalic and nongenitalic mating structures is sometimes difficult. I use the term *male genitalia* in a broad sense to include all intromittent structures on the male body as well as the packages of sperm *(spermatophores)* that males of some groups deposit outside their bodies and then use instead of or in addition to intromittent organs. In many species males have additional structures *(claspers)* for holding the female during copulation. Standard usage includes some but not all of these in the term genitalia, the critical factor being the distance from the male gonopore. For the moment I will include as genitalia all male structures that are inserted in the female or that hold her near her gonopore during sperm transfer. In Chapters 11 and 12 I will discuss why this decision is arbitrary and in fact misleading, but it will serve for the moment as a working definition.

I include the male reproductive organs that are directly involved with copulation but I exclude others, such as testes, accessory glands, and so on that are not normally in direct contact with females. Roughly speaking, I include external but not internal male reproductive structures. The reason for this distinction is that the structures included are those that tend to evolve rapidly and divergently. In sum, all male structures directly associated with the male gonopore or specialized for intromission or for holding the female near her copulatory pore will be called genitalia.

Female genitalic structures involved in copulation are more consistently associated with the copulatory opening and thus are easier to characterize. It is nearly universally true, however, that male intromittent structures reach only to the "lower" parts of the female reproductive tract and so do not deposit sperm directly onto the female's eggs. (Evolutionary consequences of this fact are discussed in Chapter 6.) In keeping with the definition above, I will consider as genitalia those parts of the female reproductive tract that make direct contact with male genitalia or male products (sperm, spermatophores) during or immediately following copulation. Specifically excluded are those structures higher up, such as ovaries, and others not in direct contact, such as accessory glands, shell glands, and such.

Rapid and divergent. The fact that taxonomists can often find greater differences between related species in genitalia than in other structures implies that relative to the other structures the genitalia have diverged rapidly. It says very little about the actual rates of divergence (amount of change per year), since some species are undoubtedly quite old while others are very young. The terms *rapid* and *divergent* are used throughout in a comparative rather than an absolute sense.

A survey of animal genitalia

With these definitions in mind we can proceed to survey the evolution of genitalia in different animal groups. The data are given in Tables 1.1 – 1.4, and a number of specific cases are illustrated in Figs. 1.1 – 1.8. Only male genitalia are shown, since female genitalia commonly seem to vary less (see Chapter 3), and they are less often described. Table 1.1 lists nineteen major groups in which male primary genitalia (copulatory structures located close to the male gonopore) are used as taxonomic characters to distinguish species. Table 1.2 lists fourteen groups in which males do not copulate with their primary genitalia but instead use other body parts (secondary genitalia) as intromittent organs. In these groups the secondary male genitalia have evolved rapidly and divergently while the primary genitalia have remained relatively simple and uniform. Table 1.3 lists fifteen groups in which the male produces a spermatophore that is inserted into the female, and in which the shape of the spermatophore has evolved rapidly and divergently. Again in these cases the primary male genitalia tend to be relatively uniform. Table 1.4 lists eight groups in which fertilization is external and four others in which males produce simple spermatorphores. Male genitalia in these groups consistently fail to show species-specific variations.

The most striking aspect of the data in these tables is the extraordinary range of animals — from planarians and nematodes to snakes and rodents — that show rapid and divergent genitalic evolution. In fact no single major group in which internal fertilization is common fails to show this pattern. This consistency is especially impressive because the tables probably underestimate the diversity of genitalic form. For instance, in some animals with intromittent male genitalia, such as mammals, ratite birds, and ducks, specimens are generally preserved in a way (for example, dried skins) that makes study of male genitalic form effectively impossible. In addition, genitalic

Table 1.1 Groups in which male primary genitalia have evolved rapidly and divergently and are useful characters in species-level taxonomy. (Dash indicates no information found.)

Group	Female genitalia soft and saclike or tubelike?	Sperm deposited directly onto eggs?	References
Most insects[a]	Some	No	Jeannel 1941; Imms 1957; Tuxen 1970
Mammals[b]			
Artiodactyls[b,c]	Yes	No	Gerhardt, in Walton 1960
Bats[a]	Yes	No	Krutzch and Vaughn 1955; Martin and Schmidly 1982
Cavimorph and microtine rodents[a]	Yes	No	Walton 1960; Hooper and Musser 1964; Prasad 1974
Primates[c]	Yes	No	Short 1979; Hershkovitz 1979
Many pulmonate molluscs[a,d]	Yes	No	Baker 1945; Duncan 1975
Some opisthobranch molluscs[c]	Yes	No	Edmunds 1969, 1970
Poeciliid fish[c]	Yes	No	Rosen and Gordon 1953; Rosen and Bailey 1963
Some cottid fish[a]	Yes	No[e]	Watanabe 1960
Sharks and rays[e,f,g]	Yes	No	Leigh-Sharpe 1920, 1922; Ishiyama 1967; Applegate 1967
Many snakes[a]	Yes	No	Dowling and Savage 1960; Klauber 1972; Saint-Girons 1975

Some lizards[a,b]	Yes	No	Uzzell 1966; Cuellar 1966; Rosenberg 1967; Arnold 1973; Saint-Girons 1975; Connor and Crews 1980; Branch, in press; A. Kluge, 1982
Many nematodes[a]	Yes	No	Hyman 1951a; Hope 1974; Spratt 1979
Many turbellarian flatworms[a,d]	Yes	No	Hyman 1951a; Meglitsch 1967; Henley 1974
Some polychaete worms[a]	Yes	No	Westheide 1967; Schroeder and Hermans 1975
Many oligochaete worms[a,d]	Yes	No	Stephenson 1930; Avel 1959; Lasserre 1975
Ostracod crustaceans[a]	—	—	Pennak 1978
Some opilionids[a]	—	—	Forster 1954; Kaestner 1968
Some mites[a]	No	No	Pritchard and Baker 1955; Fain 1967, 1981; Santana 1976; Griffiths and Boczek 1977; B. OConnor, pers. comm.

a. Author states clearly that structure is useful in distinguishing species.
b. Lists probably underestimate the number of groups due to the difficulty of preparing hemipenes for study (lizards) and the lack of proper preservation of genitalia (mammals). Eckstein and Zuckerman (1956) and Prasad (1974) document the diversity (and thus probably the taxonomic usefulness) of mammalian penes.
c. In some cases genitalia are useful as genus rather than species characters.
d. Most species are hermaphrodites.
e. Small size of intromittent organ apparently precludes deposition of sperm directly onto eggs.
f. Species specificity of genitalia deduced from study of figures rather than from authors' statements. Ishiyama (1967) did use male clasper organ as identifying character in rays.
g. Male morphology suggests that some clasper organs may be used to flush out previously deposited sperm from the female.

Table 1.2 Groups with indirect sperm transfer in which secondary male genitalia have evolved rapidly and divergently and are useful characters in species-level taxonomy. (Dash indicates no information found.)

Group	Structure that transfers sperm	Sperm deposited directly onto eggs?	Primary male genitalia useful characters?	References
Rhodacaridae (mites)[a]	Chelicerae	No	No[b]	Lee 1970
Proctolaelaps (mite)[a]	Chelicerae	No	—	B. OConnor, pers. comm.
Some gamasine mites[b]	Chelicerae	—	—	Treat 1975
Piona (mite)[a]	Tarsus and claw III	—	—	Mitchell 1957
Decapod crustaceans[a]	Pleopods I and II	No	—	Andrews 1904, 1911; Pennak 1978
Most millipeds[a]	Legs of ring III	No	No[b]	Kaestner 1968; Shear 1976, 1981; Shelley 1981
Isopod crustaceans[a]	Pleopod II	—	—	Pennak 1978; Pires 1982
Odonata[a]	Abdominal sternites	No	No[c]	Kennedy 1920; Imms 1957; Corbet 1962; Tuxen 1970
Spiders[a]	Pedipalps	No	No[b]	Kaston 1948; Kaestner 1968
Ricinuleids[a]	Leg III	No	No[b]	Kaestner 1968; Brignoli 1974; Platnick and Shadab 1976, 1977
Solifuges[a]	Chelicerae	—	No[b]	Kaestner 1968; H. Levi, pers. comm.
Most cephalopod molluscs[a]	Arm(s)	No	No	Hoyle 1907; Wells and Wells 1977; Arnold and Williams-Arnold 1977; Wells 1978
Copepods[a]	Leg V	—	—	Davis 1949
Dactylochelifer (pseudoscorpion)[a]	Tarsus leg I[d]	No	No[b]	Weygoldt 1969; Sato 1982

a. Author states that structure is useful in distinguishing species.
b. Species specificity, or lack of it, of genitalic structure deduced from study of figures and/or from lack of inclusion in lists of taxonomically useful characters rather than from author's statement.
c. Clasping structures (superior appendages) near primary genitalic opening used to seize the female are often species-specific, but primary genitalia that are involved in transfer of sperm are relatively invariable.
d. Claw opens female atrium and "assists" in uptake of spermatophore.

Table 1.3 Groups in which the spermatophore has evolved rapidly and divergently and is a useful character in species-level taxonomy. (Dash indicates no information found.)

Group	Female genitalia soft and saclike or tubelike?	Sperm deposited directly onto eggs?	Spermatophore injects sperm into female?	References
Some salamanders[a]	Yes	No	No[b]	Jordan 1891; Mohr 1931; Noble and Brady 1933; Organ and Lowenthal 1963; Uzzell 1969; Spotila and Blumer 1970; Boisseau and Joly 1975
Some pseudoscorpions[c]	—	No	Yes	Weygoldt 1969
Amblypygids (arachmids)[c]	—	No	No	Alexander 1962a,b; Weygoldt et al. 1972
Uropygids (arachnids)[c]	—	No	—	Schaller 1971
Scorpions[c]	—	No	Yes	Kaestner 1968; Francke 1979
Some Hirudinea (annelids)[a,d]	Yes	—	Yes	Mann 1962; Lasserre 1975
Some lumbricid worms[a,d]	Yes	No	—	Stephenson 1930
Pogonophoran worms[a]	Yes	—	—	Southward 1975
Cephalopods[a]	Yes	No	Yes	Pickford 1945; Wells and Wells 1977; Wells 1979
Some slugs[a,d]	Yes	No	No	Quick 1947, 1960
Diodora (mollusc)[a,d]	Yes	No	—	Fretter and Graham 1964
Siphonaria (pulmonate mollusc)[a,d]	Yes	No	—	Berry 1977
Some Lepidoptera[a]	Sometimes[e]	No	No	Norris 1932; Imms 1957; Burns 1968
Some chaetognaths[d,f]	Yes	No	No	Reeve and Cosper 1974; Alvariño 1965; Ghirardelli 1968

a. Author states that structure is useful in distinguishing species.
b. Sperm mass is generally sticky.
c. Species specificity is not established, but the structural complexity of the spermatophores indicates that they probably are often species-specific.
d. Most species are hermaphrodites.
e. An internal hard structure of the female, the signa, is often species-specific.
f. The seminal vesicle acts as a spermatophore and has a species-specific form.

Table 1.4 Lack of morphological diversity in male primary genitalia of groups with external fertilization and groups in which males and females do not make contact before insemination.

Group	Male genitalia species-specific in form?	References
Echinodermata[a]	No[b]	Hyman 1955
Hemichordata[a]	No[b]	Hyman 1959
Phoronida[a]	No[b]	Hyman 1959
Brachiopoda[a]	No[b]	Hyman 1959
Sipunculida[a]	No[b]	Hyman 1959
Most Polychaeta[a]	No[c]	Hartman 1965, 1967; Fairchild 1977
Most Pisces[a]	No	Greenwood 1975
Amphipoda[a] (Crustacea)	No[e]	Bousfield 1958
Thysanura[d] (Insecta)	No[e]	Tuxen 1970
Campodeid Diplura[d] (Insecta)	No[e]	Tuxen 1970; Schaller 1971
Protura[d] (Insecta)	No[e,f]	Tuxen 1970
Collembola[d] (Insecta)	No[e]	Tuxen 1970; Schaller 1971

a. Fertilization is external.

b. Author states that sexes cannot be distinguished on the basis of external morphology.

c. List of characters used in taxonomy does not include genitalia, and descriptions of new species do not mention them.

d. Fertilization is internal but occurs by means of a spermatophore, without contact between males and females.

e. Author states that external male genitalia are simple or lacking and do not distinguish species.

f. Female genitalia are important taxonomic characters.

complexity is probably underestimated in most of the groups listed, because genitalia are usually studied in the retracted or relaxed state; evidence from groups in which genitalia have been studied in expanded form (see Chapter 9 and Fig. 9.1) suggests that they are more obviously complex and diverse when they are expanded. In some groups (for example, some mites), mature males are rare in collections, and taxonomists have for practical reasons been reluctant or unable to use male structures to distinguish species. Finally, in still other groups, including some mites, tingid bugs, membracid treehoppers, and cerambycid beetles, the relative difficulty of dissecting out and examining male genitalia, compared with the ease of using other body characters, has led taxonomists to ignore the genitalia; in some of these groups recent studies have shown that male genitalia do provide additional useful characters (see, for example, Kopp and Yonke 1973; Deitz 1975 on membracids).

In sum, rapid and divergent evolution in male genitalic structures is an extraordinarily widespread trend in animals with internal fertilization. In contrast, male genitalia in groups with external fertilization are consistently simple and uniform. The association between internal fertilization and rapid and divergent evolution of male genitalia is so strong that it suggests there must be some extremely general explanation having to do with the act of copulation. Explanations that depend on particular characteristics of a given group but are inappropriate for a large number of species in other groups are not likely to be applicable even for the groups with special characteristics. The search in this book is thus for a truly general explanation, and evidence against the generality of a hypothesis will be taken as cause for rejecting it.

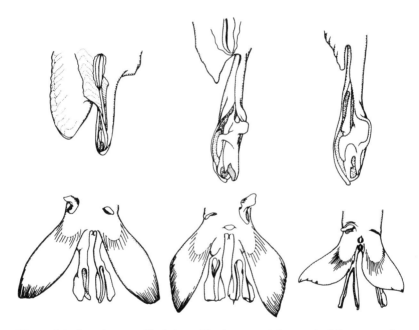

Figure 1.1 Species-specific intromittent organs (claspers) of three species each of the ray genus *Raia* and the chimerid fish genus *Chimaera*. The clasper organs of some rays are known to swell substantially during copulation. (After Ishiyama 1967.)

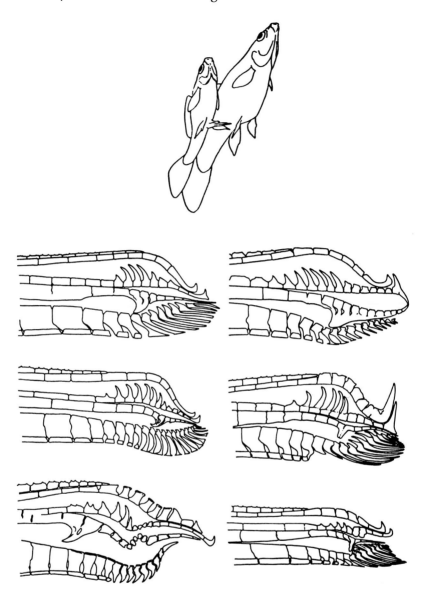

Figure 1.2 Above, internal insemination in a poeciliid fish; the male uses his modified anal fin (gonopodium) as an intromittent organ. *Below,* the bones in the gonopodium vary between species, as illustrated in six species of *Gambusia.* (Top, from Clark et al. 1954; others from Rosen and Bailey 1963.)

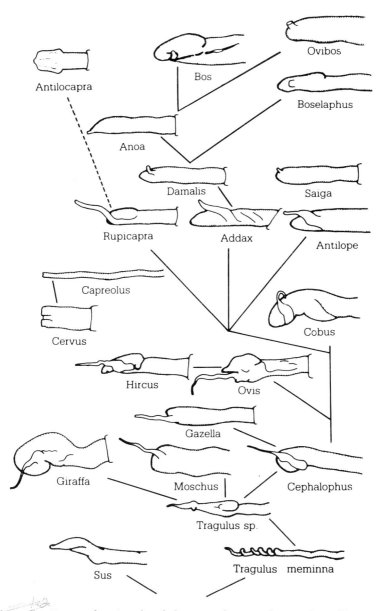

Figure 1.3 Penes of various hoofed mammals. Many have a fingerlike projection which varies between genera in size, shape, and position; observations of the bull *(Bos)* show that it flips forward inside the female at the moment of ejaculation and withdrawal and may serve as a stimulator. (After Walton 1960.)

Genitalic extravagance

The common-sense function of male genitalia is that of gamete transfer: in one way or another the male places his sperm inside the female, where they can then fertilize her eggs. But when one looks at the extraordinary complexity of the male genitalia of a species like the chicken flea in Fig. 1.9, for example, this simple explanation seems inadequate. In many species the male genitalia are structurally the most complex organs in his entire body; in some they are

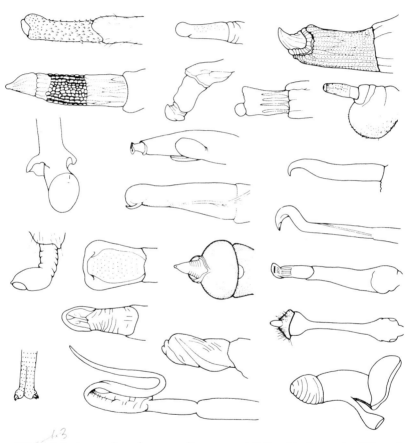

Figure 1.4 A sampler of mammalian penes (all flaccid, drawn to different scales) from groups in which penis morphology is generally poorly known and has not been used to make taxonomic distinctions. Those in the top two rows are all primates. The forms are varied and complex, suggesting that further study may reveal species-specific differences. (After Prasad 1974.)

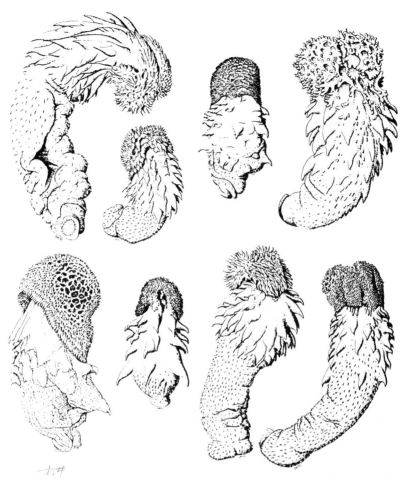

Figure 1.5   Hemipenes from eight different species of the snake genus *Rhadinaea*, inflated as they are when inside the female. Snakes have paired penes; only one side is shown in the drawings. (From Myers 1974, courtesy of the Library Services Department, American Museum of Natural History.)

also incredibly large, as in some nematodes and flies, whose intromittent organs are longer than the rest of the body (Hyman 1951a; Thornhill and Alcock 1983; see also Fig. 5.3). It is just too fantastic to believe that such complicated machinery is necessary only to perform a mechanically simple function.

In the following chapters I will discuss a number of attempts that have been made to explain this extravagance. Darwin himself provided a clear precedent in the analysis of apparently extravagant sexually dimorphic (and usually species-specific) structures. In *The Descent of Man and Selection in Relation to Sex,* he showed that structures ranging from beetle horns to bird feathers could be explained by the action of selection favoring those individuals, usually males, who are best able to compete with other members of the same sex for mates. As will be shown in later chapters, sexual selection is also the most likely explanation for genitalic extravagance. I argue that in addition to transferring sperm, male genitalia also function as "internal courtship" devices to insure that the male's

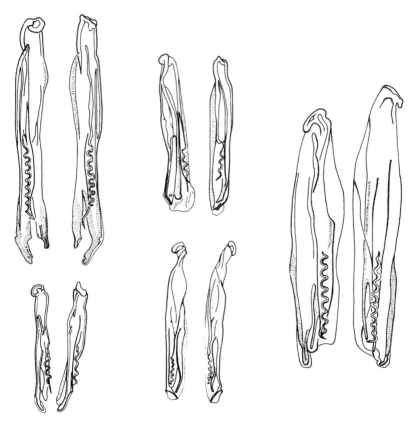

Figure 1.6 Left and right male intromittent organs (spicules) of five species of *Paracooperia* nematodes (drawn to different scales). Even such structurally simple animals as nematodes can have complex, divergent male genitalia. The spicules of these species are not only elaborate, they are also consistently asymmetrical in both size and shape. (After Gibbons 1978.)

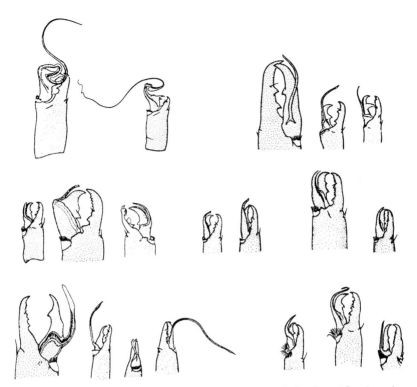

Figure 1.7 Male chelicerae of different species of rhodacarid mites, in which chelicerae are used as intromittent organs. The species in each group are in the same genus: there are clear differences both between and within genera. (From Lee 1970, reproduced courtesy of the South Australian Museum.)

sperm will be used to fertilize the female's eggs. The extraordinary complexity of male genitalia has probably evolved as a result of selection acting on this second function.

The concept of sexual selection by female choice has had a strange history. Darwin (1871) originally proposed it, along with sexual selection by male-male combat, to explain a spectacular array of secondary sexual characters that he thought could not be explained as adaptations to improve their bearers' ability to cope with their environment. So strong was Darwin's apparent belief in the powers of sexual selection that if the complexity and variety of genitalic structures had been common knowledge among zoologists in his day, or even, perhaps, if he had studied beetles rather than barnacles, I suspect he would have included genitalia in his listing.

People were reluctant to accept Darwin's claims that females possess an "aesthetic" sense with which they judge the "beauty" of males, perhaps partly because he used such anthropomorphic terms. It was also difficult to believe that females actually rejected some males because they lacked apparently trivial characters that were unrelated to the ability to grow and survive. Indeed, Darwin himself thought female choice most important in vertebrates because of their "higher faculties" of judgment and memory. Another, stronger objection was that many of the characteristics that Darwin thought had evolved to impress females are also used in aggressive interactions among males (see, for example, Borgia 1979). For this and other reasons (see West-Eberhard 1983), the idea fell into neglect.

During the revolution in evolutionary theory initiated largely by the works of Hamilton (1964) and Williams (1966), the motivating force for sexual selection — unequal investment in offspring by the

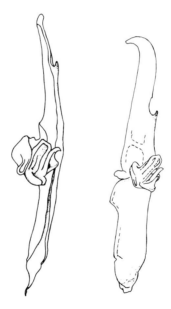

Figure 1.8 Spermatophores of two species in the scorpion genus *Opisthacanthus*, showing their complex, species-specific morphology. (From Lourenço 1980.)

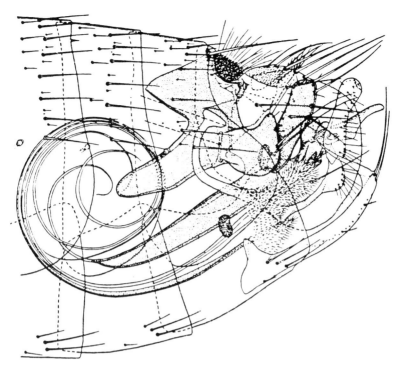

Figure 1.9 One of the marvels of organic engineering, the male genitalia of the chicken flea *Ceratophyllus gallinae.* Such morphological exuberance is difficult to explain on the assumption that the only function of genitalia is the transfer of gametes. (From Smit 1970, by permission of the Royal Entomological Society and the Trustees of the British Museum [Natural History].)

two sexes — was rediscovered (Trivers 1972; see Darwin 1871 for the original explanation of its importance). There has been a recent surge of interest in both types of sexual selection (Blum and Blum 1979; O'Donald 1980; Thornhill and Alcock 1983; Ryan, in press).

The extension of sexual selection by female choice to genitalia breaks new ground in two different respects. On the one hand, genitalia are almost never used in direct male-male combat, and I will show that they are probably used only occasionally in other direct male-male competitive interactions; the old objection that male combat characters are being incorrectly analyzed does not generally apply and genitalia thus may be unusually "pure" products of sexual selection by female choice. On the other hand, female discrimination is attributed to much simpler animals, such as planarians and nematodes.* The existence of discriminatory abilities in such animals is not as controversial now, however, as it would have been in Darwin's time. Behavioral studies have now shown that even very simple organisms are able to sense and respond to stimuli such as those produced by species-specific genitalic structures, and responsiveness to simple male-produced stimuli, rather than a complex "aesthetic" sense, can result in sexual selection by female choice (see also West-Eberhard 1984).

---

* In fact Darwin, in response to a critic of sexual selection, specifically denied that the clasping organs (bursae) of male "parasitic worms" (probably nematodes) resulted from sexual selection by female choice (Darwin 1871, p. 569 and footnote 4).

2   *Previous hypotheses: their status to date*

Despite the generality of rapid and divergent evolution in genitalia, there has been surprisingly little research on the reasons for it. The reason for this neglect might make an interesting study in the sociology of science. My impression is that animal taxonomists, who have been most directly concerned with the evolution of genitalia, are usually empirical rather than theoretical in orientation. In any case, I know of only four hypotheses explaining why genitalia change both rapidly and divergently. In this chapter I will present each of them briefly and give a historical overview of the evidence that has accumulated up until now both for and against them. In the chapters that follow I will give additional arguments against them.

Lock and key

This explanation is the oldest (Dufour 1844, in Mayr 1963), and the most often invoked, even though a number of studies have shown that it is unlikely to be true for many groups (Richards 1927a; Hewer 1934; Fennah 1945; Gordon and Rosen 1951; Lorkovic 1952; Gering 1953; Clark et al. 1954; Jeannel 1955; Kunze 1959; Alexander and Moore 1962; DeWilde 1964; Klauber 1972; Bacheler and Habeck 1974; Waage in press; see also references in Dobzhansky 1941 and Mayr 1963). According to this theory females avoid having their eggs fertilized by the males of other species by evolving complicated genitalia that permit insemination only by the corresponding genitalia of males of their own species; the male has the "key" to fit the female's "lock." Each new speciation event necessitates a new lock and a new key.

Most of the evidence that has been presented both for and against this hypothesis is weak. Some studies that are usually taken to provide evidence against mechanical isolation (for example, Richards 1927a; Jeannel 1941; Lorkovic 1952; Gering 1953; Kunze 1959; Paulson 1974 on *Enallagma carunculatum* and *E. boreale*) have been limited to showing that it is possible for males of other species to effect mechanical coupling, without documenting whether or not normal amounts of sperm are passed and used in fertilization. Other studies showed that interspecific fertilizations can occur despite genitalic differences (for example, Beheim 1942, in Mayr 1963), but have not documented whether normal amounts of sperm were passed and survived or if subsequent female receptivity, or nonreceptivity, was normal. Since females could benefit from partial as well as absolute discrimination against cross-specific matings, these factors must be taken into consideration in testing the lock and key hypothesis.

Hand-in-glove fits between male and female genitalia have been cited as evidence favoring the lock and key argument (see Pope 1941; Watson 1966; Coe 1969; Fooden 1970; Hafez 1973; Freitag 1974; Toro and de la Hoz 1976). Figure 2.1 gives a mechanical interpretation of such a fit in a spider. But this support is weak because other hypotheses, including that of sexual selection by female choice, also predict close correspondence between male and female structures in at least some cases. Even demonstrations that interspecific differences in genitalia or clasping structures do make it difficult or impossible for different present-day species to couple (Paulson 1974; Hopkins, in Gagné and Peterson 1982) are not sufficient to show that reproductive isolation was the selective factor responsible for the divergent evolution in the first place (Robertson and Paterson 1982). It is quite possible that divergence caused by other factors sometimes has the incidental effect of making cross-specific matings more difficult.

Perhaps the most convincing data that have been cited in favor of the lock and key hypothesis (they also favor the genitalic recognition hypothesis — see next section) is that some taxonomic groups whose courtship behavior is particularly elaborate have comparatively simple and uniform genitalia, compared with related groups that have less elaborate courtship (Rentz 1972; see Table 2.1 and Fig. 2.2). This is in accord with both the lock and key and genitalic

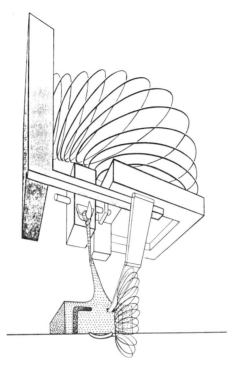

Figure 2.1 A mechanical interpretation of the mesh between the male pedipalp (intromittent organ) and the female epigynum (stippled) of an *Araneus* spider. The pedipalp's complex set of springs, braces, and catches emphasizes its Rube Goldberg-like structure. Mechanical meshes like this, which depend on close fit between male and female structures, are common in some groups but are by no means universal. (After Grasshoff 1973.)

recognition hypotheses, because courtship behavior probably serves as a preliminary filter of possible mates, and "incorrect," cross-specific pairings resulting in genitalic contact probably occur less often in species with more complex preliminary behavior.

The support is not convincing however, partly because it is difficult to determine just what cues animals use to sense species identity. The fact that cues are available does not necessarily mean they are used (see Silberglied and Taylor 1978 on pierid butterflies). In addition, species whose courtship seems to be simple or to be lacking altogether may actually engage in complex, species-specific behavior that occurs too quickly for the unaided human eye to perceive (see Tobin and Stoffolano in Ewing 1977 on houseflies,

Table 2.1 Pairs of groups showing correlation between complex, species-specific courtship behavior and simple, uniform male genitalia. The list is undoubtedly incomplete.

Group	Courtship	Genitalia species-specific?	References
Slant-face and band-wing grasshoppers	Complex (visual and auditory)	No	Otte 1970; Rentz 1972
Spur-throat grasshoppers	Simple[a]	Yes	Otte 1970; Rentz 1972
Some North American odonates	Complex (visual)[b]	No[c,d]	Paulson 1974
Other North American odonates	Simple[b]	Yes[d]	Paulson 1974
Hawaiian Drosophilidae (drosophiloid group)	Complex (visual, chemical, auditory)	No[e]	Hardy 1965; Kaneshiro 1983
Hawaiian Drosophilidae (scaptomyzoid group)	Simple	Yes	Hardy 1965; Kaneshiro 1983
Oncopis leafhoppers	Complex (auditory)	No	Claridge and Reynolds 1973
Many other leafhoppers	Simple(?)[f]	Yes	Claridge and Reynolds 1973
Tibicen and *Magicicada* cicadas	Complex (auditory)	No[g]	T. Moore, pers. comm.
Other cicadas	Simple	Yes	T. Moore, pers. comm.
Some *Desmoprachia* water beetles	Complex (visual)[h]	No	Young 1981; Weems 1981
Other *Desmoprachia*	Simple[i]	Yes	Young 1981; Weems 1981
Some megachilid bees	Complex (tactile)[j]	No	Griswold 1983; G. Eickwort, pers. comm.
Other megachilids	Simple[k]	Yes	Griswold 1983; G. Eickwort, pers. comm.

Table 2.1 *(continued)*

Group	Courtship	Genitalia species-specific?	References
Rhinoseius (mites)	Complex (tactile)[l]	No	B. OConnor, pers. comm.
Proctolaelaps (mites)	Simple[m]	Yes	B.OConnor, pers. comm.

a. Correlation is incomplete; courtship of *Microtylopteryx* grasshoppers is relatively complex, but male genitalia are important characters for distinguishing species (E. Braker, pers. comm.).

b. Paulson judged the likelihood of visual species isolation on the degree of interspecific differences in female color in sympatric congeneric species.

c. Male abdominal clasping structures, not secondary genitalia, used in identification.

d. Correlation is incomplete; four genera have interspecific differences in female color but have species-specific male clasping structures, and *Argia* also has species-specific intromittent organs (Kennedy 1919; see Fig. 2.3).

e. Correlation is incomplete; drosophiloid male genitalia are less complex and less distinctive, but they are nevertheless key taxonomic characters and are the *only* characters known to separate some pairs of species (Hardy 1965).

f. Evidence is weak; males of other genera also may produce species-specific songs (see *Dalbulus* in Table 2.2), but this has not been adequately investigated.

g. Some differences between species occur but are less clear than those in other groups.

h. Elytral colors are bright; courtship cues have not been well studied.

i. Elytral colors are dull; courtship cues have not been well studied.

j. Male legs and sterna, which contact females, are species-specific in form (see Fig. 2.2).

k. Male legs and sterna are not species-specific in form.

l. Female mounts male before mating and may contact species-specific male dorsal setae.

m. Female apparently does not mount male before mating.

Musca). Or they may use signals such as chemicals that are not easily detected by unaided human observers.

In addition, the general correlation does not hold in some groups (Table 2.2). Perhaps the most dramatic case is that of the damselfly genus *Argia* in which there are both interspecific color differences, which probably reduce the chances of cross-specific pairs ever forming (Paulson 1974), and male clasper differences, which reduce the chances of males successfully seizing females of other species (Paulson 1974) and probably also reduce the males' chances of inducing females to copulate (see Robertson and Paterson 1982). Nevertheless, the penis morphology is generally species-specific, as shown in Fig. 2.3 (Kennedy 1919). (L. K. Gloyd has pointed out to me that there are some exceptions, but even in these cases penis morphology is consistently specific within groups of species.) There are also exceptions to the supposed rule of complex courtship correlating with simple genitalia in some of the groups that appear to show

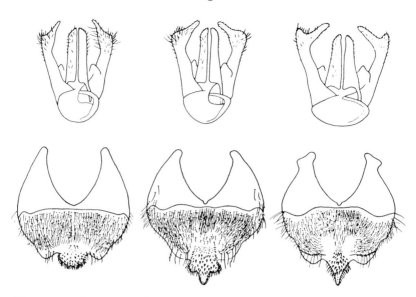

Figure 2.2 Above, genitalia and, *below,* abdominal sterna of males of three species of megachilid bees of the genus *Acrosmia.* Here the apparent complexity of precopulatory signals (the male sternum presumably stimulates the female when he mounts her) is associated with relative uniformity in genitalia, as predicted by both of the species isolation hypotheses. (From Griswold 1983.)

the correlation (for example, Hawaiian drosophilid flies and grasshoppers; see footnotes a, d, and e in Table 2.1). Finally, a mixture of correlation and lack of correlation is predicted by the sexual selection hypothesis (see Chapter 7), so the support for species isolation is even less convincing.

A strong theoretical reason to doubt the two reproductive isolation hypotheses (lock and key and genitalic recognition) has been noted by Alexander (1962, 1964) and McGill (1977; see also references in Alexander and Otte 1967). Natural selection should favor females who are able to determine males' species identity early in courtship sequences rather than late, since both courtship and copulation are often somewhat costly and dangerous for a female (Daly 1978). Thus species discrimination by means of genitalia, while feasible, would be less advantageous than discrimination based on stimuli received prior to copulation. Even if some females do use their genitalia to identify males of the same species, discrimination based on stimuli received prior to copulation would be

Table 2.2 Groups showing lack of correlation between complex, species-specific courtship behavior and simple, uniform male genitalia. The list is undoubtedly incomplete.

Group	Courtship	Genitalia species-specific?	References
Pteropteryx and *Pyractomena* fireflies	Complex (visual)	Yes	Green 1957; Ballantyne and McLean 1970; Lloyd 1983
Utetheisa moths	Complex (chemical)	Yes	Connor et al. 1981; J. Franclemont, pers. comm.
Lipara gall flies	Complex (vibratory)	Yes	Chvala et al. 1974
Hilara dance flies	Complex (visual)	Yes	Chvala 1971; Ewing 1977
Sarpogon robber flies	Complex (visual)[a]	Yes	Theodor 1980
Many damselflies and dragonflies *(Argia, Enallagma, Archilestes)*	Complex (visual, mechanical)	Yes[b]	Kennedy 1919, 1920; Paulson 1974; Garrison 1982; Robertson and Paterson 1982
Dalbulus leafhoppers	Complex (auditory)[c]	Yes	Nault 1984; L. R. Nault and S. E. Heady, pers. comm.
Some *Meloe* beetles	Complex (tactile)	Yes	Pinto and Selander 1970
Several families of pseudoscorpions	Complex (tactile, chemical)	Yes[d]	Weygoldt 1970
Uca fiddler crabs	Complex (visual, auditory)	Yes	Crane 1975; Weygoldt 1977
Some rodents	Complex (chemical)[e]	Yes	Hooper and Musser 1964; Doty 1974; Prasad 1974

a. Complexity of courtship deduced from species-specific sexual dimorphism in color.

b. Intergeneric differences in male secondary genitalia are widespread; intrageneric differences have been documented in some groups, such as the genera cited.

c. The only congeneric species that are sympatric belong to different species groups, and male calls, which attract females, differ widely between different species groups.

d. Male spermatophore is structurally complex and probably species-specific.

e. Behavioral species isolation is probable, but few direct observations have been made.

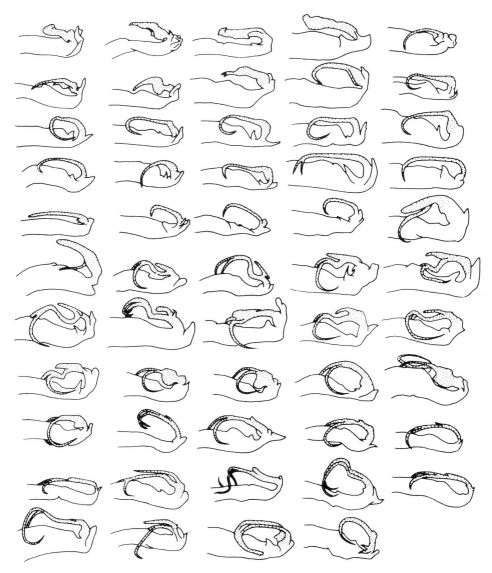

Figure 2.3 Penes of different species of the damselfly genus *Argia*. In those species that have been studied, the stippled parts are used to scoop out sperm deposited in the female by previous mates. In many species of this genus both color and the male clasping organ are species-specific (Paulson 1974), and thus it is thought that cross-specific pairs rarely, if ever, initiate genitalic contact. Even so, male penis structure is generally species-specific, contrary to the species isolation hypotheses. (After Kennedy 1919.)

favored, and thus selection on genitalia in a species isolation context would be expected to be only transitory at best.

Although the logic of this argument is clear, we cannot confidently reject the species recognition hypotheses on this account, since cross-specific pairs do sometimes form despite differences in courtship signals. (See Bick and Bick 1981 on odonates; Chaplin 1973, and Leslie and Dingle 1983, on lygaeid bugs; Frechin 1969, and Brown 1981, on butterflies; Berlese 1925, on other insects; Noble, in Pope 1941, on snakes.) Such pairings are probably infrequent in natural conditions, but few (if any) data are available to prove it. Thus it can be argued that genitalic differences may function as backup or fail-safe devices to ensure species isolation even when behavioral devices fail.

Genitalic recognition

Many authors have observed directly or deduced from morphology that male genitalia are used in ways apparently designed to stimulate females (see Stephenson 1930; Baker 1945; Clark and Aronson 1951; Lorkovic 1952; Alexander 1959; Clements 1963; Fretter and Graham 1964; Long and Frank 1968; Diamond 1970; Ewer 1973; McGill 1977; Scott 1978; Platt 1978). Some (Jeannel 1942, 1955; De-Wilde 1964; Diamond 1970; Hammond 1981) have hypothesized that the female determines species identity on the basis of species-specific genitalic stimuli and avoids fertilization if the stimuli are not appropriate. This is similar to the lock and key idea except that the criteria are stimulatory rather than mechanical. The data and arguments relating to the lock and key hypothesis apply equally to this hypothesis, giving partial but inconclusive reason to doubt that it is generally important.

Pleiotropism

Mayr (1963) proposed that genitalia are pleiotropically affected by many genes and that "any change in the genetic constitution of the species may result in an incidental change in the structure of the genitalia. As internal structures, they are less subject to the corrective influences of natural selection . . . provided the basic function of gamete transfer is not impaired" (p. 104).

This hypothesis has the apparently fatal weakness of not explaining why only genitalia and not other "internal" organs should be consistently affected by pleiotropisms (see Chapter 4). Mayr's idea was later modified by Arnold (1973) in an attempt to overcome these objections. Arnold pointed out that when a new allele that is favored by natural selection in one context also modifies another character (such as hair color or bone form), there will be subsequent selection on the rest of the genome to suppress these pleiotropic effects whenever the original hair color or bone form was adaptive, as is presumably the usual case.

But if such a favored allele pleiotropically affects the genitalia of one sex, a different situation results. Assuming the necessity of a close fit between male and female genitalia, selection will not only act to suppress the pleiotropic effects but also to modify the genitalia of the opposite sex to adapt them to the new form of the genitalia of the first sex. Often the compensatory modifications in the second sex will reduce and eliminate the selection to suppress the original pleiotropic effects. The genitalia of both sexes are thus expected to change in concert and to evolve more rapidly due to pleiotropic effects than other characters.

The status of the pleiotropism argument is not clear. On the one hand it seems improbable; evidence of the extraordinary effects of natural selection on many other body structures makes it seem doubtful that such intricate modifications as those of some genitalia could have arisen despite their having no adaptive value at all. In fact, since natural selection focuses on differences in reproductive ability, it seems probable that the male's ability to place sperm in an advantageous site within the female would be especially likely to be subject to selection. But the pleiotropism theory seems to be irrefutable by direct evidence (see, however, Chapter 4). To my knowledge there is not a single species in which the genetic control of genitalic morphology is well understood (Gordon and Rosen 1951, Turner et al. 1961, and Peck 1983 give partial accounts for fish, butterflies, and beetles), and modern geneticists seem to be convinced that pleiotropy is very common (see, for example, Falconer 1981). In addition, it is very difficult to document all the effects of a given allele, and it is also often difficult to demonstrate the precise effects of an allele on its bearer's fitness in nature. The current state of the evidence on this front is thus an uneasy standoff between biological "intuition," which suggests that it is incorrect, and an almost complete absence of concrete data either for or against.

Mechanical conflict of interest

Recent advances in evolutionary theory have clarified and extended Darwin's (1871) original insight that males and females of a given species do not always have the same reproductive interests (see Trivers 1972; Parker 1979). Several authors have proposed that male and female genitalia are in coevolutionary races against each other (R. D. Alexander, manuscript; Lloyd 1979; Wing 1982). Alexander, for example, noted that the genitalia of male orthopterans have holding organs and that if the male used these organs to force the female to do something against her best reproductive interests (such as preventing her from mating with other males, and thus also preventing her from ovipositing, foraging, and so forth), then selection would favor females who could escape from the male's grasp. Males who could overcome the females' new adaptations would then be favored in turn. Lloyd (1979) cited the possibility of arms races, in which males evolved "snippers, levers, syringes, etc." to get sperm past female resistance and into preferred sites, but he did not clarify why it would be advantageous for the female to erect such barriers. Wing (1982) suggested that male genitalia might evolve to damage female copulatory structures and thus prevent further copulations, and that females would then evolve protection against such damage. The mechanical conflict of interest hypothesis was proposed only recently, and to my knowledge it has not been extensively worked out, and no careful tests of it have yet been published.

Summary

I think it is fair to say that most people who have pondered these explanations of genitalic evolution are not particularly enthusiastic about any of them (for example, Scudder 1971). Even the data that supposedly favor one or another are in most cases seriously flawed. But in the absence of other explanations, people have given one or another of these hypotheses lip service when the general topic is raised. Taken as a whole, the data at present make the first three hypotheses relatively unappealing but still more or less tenable, while the fourth is still untested. In the following chapters I will present new arguments that I think are strong enough to eliminate all four hypotheses as general explanations.

3 Tests of the lock and key and genitalic recognition hypotheses

Lock and key

I have already summarized several kinds of evidence against the lock and key hypothesis in Chapter 2. In addition, some relatively general but hitherto unappreciated trends are not in accord with that hypothesis. In many animal groups, female genitalia are relatively uniform while the male genitalia are diverse and species-specific. These groups include (the list is undoubtedly incomplete): turbellarians, nematodes, and oligochaete worms in general (see Table 1.2); at least some genera of bumblebees (Richards 1927a); beetles (Franz, in Goldschmidt 1940; Wooldridge 1969; Hammond 1981); empidine flies (Collin 1961); lepidopterans (Norris 1932; Hewer 1934; Jordan 1905, in Goldschmidt 1940); trichopterans (Morse 1972); mecopterans (Byers 1970); homopterans (Ossiannilsson et al. 1970; DeLong and Freytag 1972a, b); morabine orthopterans (Key 1981); mallophagan lice (Ansari 1956); spiders (Milledge 1980, 1981a, b); millipeds (Kraus 1968; Shear 1972, 1976, 1981, and personal communication); slugs (Quick 1947, 1960); and planorbid snails (Baker 1945). A random check of four volumes of an entomology journal (volumes 42 – 45 of the *Journal of the Kansas Entomological Society*) confirmed that in insects this pattern is very common. Five papers stated explicitly that the genitalia of different species of congeneric females could not be distinguished, but that those of males could (Woodridge 1969, on Coleoptera; DeLong and Freytag 1972a, b, on Homoptera; Byers 1970, on Mecoptera; Morse 1972, on Trichoptera). In many additional groups male, but not female, genitalia were figured as species-specific characters, (in thirty-five of fifty-seven taxonomic papers on species in nine orders), suggesting that it is common for female genitalia to be more uniform than those of males. Scudder (1971) noted that this is true for many insect groups.

Because of technical difficulties, in part, female genitalia may not have been well studied in some of these groups; they are basically invaginated rather than everted structures and thus are more difficult to see. One might argue that the apparent lack of differences among them is an artifact of our ignorance. This criticism does not apply to many of the animals listed, however. In some groups, such as turbellarians, both males and females are studied by serial sectioning. Some (nematodes, lice, spiders) are mounted whole or as genitalic preparations in the form of semitransparent specimens on slides and studied with the compound microscope. In some groups (oligochaetes, bumblebees) the species-specific male structures contact the female's external surfaces, where modifications, if there were any, would be easily visible. And some species have been studied with the specific object of discovering female differences and have failed to find them (Kraus 1968, on millipeds; Key 1981, on grasshoppers).

It is also possible that during copulation, apparently simple female structures assume more complex configurations through pressure changes or muscle contractions. However, this is also true for male organs, which are nearly always studied in their resting state (see Chapter 9), so the relative difference between the sexes remains unexplained.

It is also worth noting that male claspers and other nongenitalic contact organs evolve just as rapidly and divergently as genitalia (see Chapter 13 and Table 13.1); the corresponding female structures in these cases are easily observed, are in most cases rigid, and are usually quite uniform. The lack of parallel changes in both male and female genitalia and nongenitalic contact organs is in clear disagreement with the lock and key hypothesis, and since this pattern is apparently common, it is strong evidence that this hypothesis cannot be a general explanation for genitalic evolution.

An even stronger argument against the lock and key hypothesis comes from the data in Tables 1.1–1.3. In many groups, including rodents and probably other mammals, snakes, lizards, poeciliid fish, sharks and rays, nematodes, turbellarians, polychaetes, oligochaetes, cephalopods, some snails and slugs, chaetognaths, and pogonophoran worms, male mating structures have evolved rapidly and divergently, even though the female genitalia are apparently

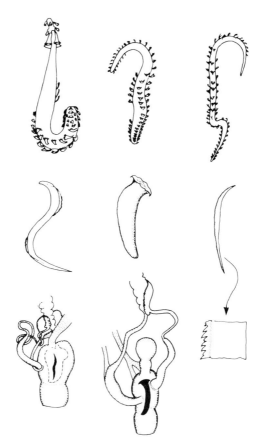

Figure 3.1 Genitalia of slugs, illustrating the improbability of the lock and key hypothesis. Spermatophores of three species are shown from the genera *Milax (above)* and *Arion (middle)*. *Bottom row* (left and middle), female *Arion* reproductive tracts with spermatophores (black) in place: there is no hint of mechanical correspondence between the shapes of male and female structures, and the female structures are soft and incapable of mechanically excluding the spermatophores of other congeneric species. *Lower right*, sawtooth edge of the spermatophore of the *Arion* species directly above. (From Quick 1947, 1960; reprinted by courtesy of *Journal of Conchology* and the British Museum [Natural History].)

simple and soft and usually cannot, except in cases of radical differences of size, mechanically exclude the intromittent structures of males of closely related species. This pattern is typified by the slug spermatophores shown in Fig. 3.1. In some species there is a close fit between male and female structures (see Clough 1969, on warthogs; Connor and Crews 1980, on an *Anolis* lizard; Pope 1941, on snakes; Hafez 1973; Fooden 1970, on primates and other mammals), but this does not constitute proof of the lock and key idea. The hypothesis depends on the female's ability to mechanically *exclude* the intromittent organs of incorrect males from sites where their sperm will have a chance to fertilize eggs, not just on a more or less precise fit between the male and female genitalia. In some of these groups the male's intromittent structure in fact assumes its species-specific, fully inflated shape only after being inserted into the female (Edgren 1953 on snakes) and thus is undoubtedly molded by the female's form. The lock and key idea was originally proposed for insects, in some species of which female genitalia are more or less rigid and thus might be able to mechanically exclude incorrectly shaped male genitalia. But this hypothesis cannot explain the same pattern of evolution in many other groups. It is probably not applicable as a general explanation for insects either; careful studies have shown that in some insect species the female genitalia probably are not capable of mechanical exclusion (Jeannel 1941).

Species isolation

In Chapter 2 I noted the inefficiency of using genitalia to exclude incorrect species, compared to the use of precopulatory recognition signals, and I concluded that for most species the selective context in which genitalia could be used for species recognition is probably fairly rare. Both the lock and key hypothesis and the genitalic recognition hypothesis also predict certain patterns of genitalic differences among species that are logical consequences of the supposed species-isolation function of the genitalia. Thus the two hypotheses can be tested empirically, and I will show that the evidence does not support the predictions.

If either hypothesis is correct, then the degree of genitalic divergence that evolves should correlate with the frequency of potential reproductive contacts with individuals of closely related species. Thus one prediction is that the genitalia of closely related species should tend to be more different in zones of sympatry than in zones of allopatry (that is, they should often show character displacement, Brown and Wilson 1956). This prediction is difficult to test, however. Present patterns of sympatry and allopatry are probably not reliable indicators of past biogeographic relationships, because the ranges of many species have changed drastically and repeatedly in the recent evolutionary past (Brown et al. 1974; Coope 1979). Also, some (possibly many) species that are thought to be allopatric may actually be parapatric or sympatric in areas that have not yet been adequately collected (Key 1981). This uncertainty regarding geographic ranges holds even for apparently strictly isolated groups, such as cave species that are endemic to different caves; for evidence that isolation is not always complete, see Deeleman-Reinhold and Deeleman (1980) on cave organisms from Yugoslavia; and S. B. Peck (in prep.) on cave organisms in the southeastern United States. It would thus be difficult to collect convincing data relating to the character displacement prediction.

More convincing data can be obtained for a related prediction. Species that have apparently been isolated from contact with all near relatives ever since their original reproductive separation should tend not to show the typical genitalic differentiation seen in nonisolated species, since the frequency of possible interspecific pairings in isolated species will be very close to zero.

The details of the predictions are best illustrated by a hypothetical example. Imagine a species living in a continental population where it is sympatric with other species of the same genus. A few individuals then colonize a distant island that is uninhabited by any closely related species. Subsequent genitalic differentiation in the new island population should be slow relative to the differentiation of other characters, because there would be no selective pressure favoring genitalic change, and because somatic characters would often change relatively rapidly because of the different biotic environment on the island. Genitalic differentiation in the isolated population should thus lag behind genitalic differentiation in other, nonisolated relatives. Over time, and with increasing differentiation in other body characters (marked taxonomically by the classification of the isolated population in a separate species group or an endemic genus), the genitalia of the isolated species should gradually lose structures that had formerly been used as isolating devices, becoming simpler and simpler. For example, mechanical barriers or neural discriminatory abilities that formerly functioned to impede cross-specific matings would no longer be favored and would gradually be lost.

Thus two concrete predictions are made for such isolated species: depending on the length of time a population has been isolated, (1) its genitalia should be less distinctive than is usual for other, nonisolated species of the same taxonomic group that have the same degree of difference in other somatic characters (that is, the genitalia should be less useful as species characters); and (2) the genitalia should be structurally simpler than those of nonisolated species.

Two kinds of strict isolation have probably often persisted unbroken since the onset of reproductive isolation and thus provide test cases for the predictions: endemic species on isolated islands and host-specific parasites that mate only in or on host species harboring no other closely related parasites. By using strict criteria to judge a species' isolation — no other species of the same genus is known from the same host species or island — and by using only well-known groups that have been recently revised and are found in well-collected areas or hosts, one can compile a list of species whose unbroken isolation from all close relatives is fairly certain. Since new species that lacked genitalic differences, as predicted by the hypotheses, might be overlooked by taxonomists, one must also eliminate from consideration those groups in which no characters other than genitalia have been used to discriminate species. Data from taxonomic papers, including statements by the taxonomists themselves and/or their drawings constitutes the evidence for the test cases.

Species on geographic islands

The taxonomic literature on island organisms is vast. Instead of attempting a general review, I concentrated on two areas, the Galapagos Islands and the subantarctic islands near New Zealand. Both areas consist of a small number of islands and have received considerable attention from biologists. In addition, the aubantarctic islands are ecologically simple (Gressitt 1962; Gressitt and Wise 1971), so our present knowledge of their faunas and their distributions may be reasonably complete. A few examples from other sites are also included.

I found twenty-four genera and one family in which two or more species are endemic to different islands within a geographic area, and are isolated on these islands from all congeners; their genitalia do not tend to be less distinct or simpler compared with those of other species of these genera as predicted (Table 3.1). Even groups with apparently ancient isolation, such as the four endemic genera of rhaphidophorid crickets (Richards 1971a, b; 1974) and *Atlantea* and *Antillea* butterflies, have unreduced and species-specific genitalia; in fact, genitalic differences among the species of *Antillea* are *greater* than expected on the basis of differences in other body characters (Higgins 1981). In genera with both isolated and noniso-

Table 3.1 Genera in which more than one endemic island species or subspecies is apparently isolated from all congeners and in which nongenitalic characters are used to distinguish nonisolated species. Simplicity of genitalia was determined by comparison with those of nonisolated species. Numbers in parentheses are number of isolated endemic species/total number of species in island group. (SA = subantarctic islands; G = Galapagos Islands, WI = West Indies.)

Taxonomic group	Site	Genitalia		References
		Species-specific?	Relatively simple?	
Diptera				
Chironomus (2/2)	SA	Yes	?	Sublette and Wirth 1980
Dasyhela (2/2)	G	Yes	?	Wirth 1969
Gigantotheca (3/3)	G	Yes	?	Lopes 1978
Hemiptera				
Cyrtopeltis (2/6)	G	Yes	No[a]	Carvalho and Gagné 1968
Homoptera				
Oliarus (5/5)	G	Yes	?	Fennah 1967
Nesosydyne (4/7)	G	Yes	No[a] (at least 2 spp.)	Fennah 1967
Philatis (7/19)	G	Yes	?	Fennah 1967
Lepidoptera				
Reductocerces (3/3)	SA	Yes	?	Dugdale 1971
Xylophanes (2/2)	G	No	?	Kernbach 1962
Atlantea[b] (3/3)	WI	Yes	No[a,c]	Higgins 1981
Antillea (2/2)	WI	Yes	No[a,c]	Higgins 1981
Irenicodes[d,e,f] (1/3)	SA	Yes	Yes[c]	Dugdale 1971
Asaphodes (2/2)	SA	Yes	? No[c]	Dugdale 1971
Tinearupa[b,d,e] (1/1)[g]	SA	Yes	No[c]	Dugdale 1971
Planotortrix (1/1)[g]	SA	Yes	?	Dugdale 1971
Orthoptera				
Halmenus[b,h] (4/4)	G	Yes	No[a]	Dirsh 1969
Rhaphidophoridae (4/4)[i]	SA	Yes	No[a,c]	Richards 1970; 1971a, b; 1974

Table 3.1   (*continued*)

Taxonomic group	Site	Genitalia		References
		Species-specific?	Relatively simple?	
Coleoptera				
Namostygnus[f] (2/2)	SA	Yes	?	Ordnish 1974
Endeius (2/2)	G	Yes	?	Coiffait 1981
Diochus (2/5)	G	Yes	No[a]	Coiffait 1981
Epichorius[f] (2/4)	SA	Yes	No[a]	Watt 1971
Meropathus (3/3)	SA	Yes	?	Ordnish 1971
Pseudhelops (1/3)[j]	SA	No	?	Watt 1971
Araneae				
Cesonia (6/10)	WI	Yes	No[c]	Platnick and Shadab 1980
Mammalia				
Oryzomys[f] (1/?)	G	Yes	No[a,c]	Patton and Hafner, in press

a. Relative complexity of genitalia determined by my inspection of figures in the reference cited.

b. Genus is endemic to islands.

c. Author cited explicitly stated this.

d. All species are brachypterous.

e. Isolated endemic races can be discriminated *only* on the basis of genitalia.

f. No other members of the same family are found on islands where isolated species occur.

g. Subspecies endemic to different islands have distinct genitalia.

h. Male genitalia are more similar to those of closest relatives *(Schistocerca)* than are other characters.

i. Each species is in a monotypic genus, and no other members of the same order are found on the same island for three of the four species.

j. There are also three isolated endemic races, none of which has recognizably distinct genitalia.

lated species on the same island groups (for example, *Irenicodes* moths, *Epichorius* beetles, most Galapagos hemipterans and homopterans; see Fig. 3.2), isolated species fail to show greater genitalic differentiation between sympatric species. And in many other groups congeneric island species show clear genitalic differentiation. Of the twenty-five higher taxa listed in Table 3.1, only three (the lepidopterans *Irenicodes* and *Xylophanes,* and the beetles *Pseudhelops*) conform at all to the predictions. In *Irenicodes* the male genitalia are partially simplified, and species of *Xylophanes* and *Pseudhelops* show unusually small genitalic differences, com-

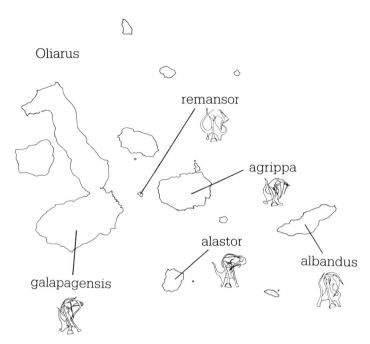

Figure 3.2 Divergent male genitalia of *Oliarus* (Homoptera, Cixiidae) species endemic to the Galapagos Islands. Only one other species of this entire family is known from the Galapagos; it is in a different genus and is sympatric with only *O. agrippa* and *O. galapagensis*. The complex, species-specific genitalic morphologies of species that have probably been isolated from all near relatives throughout their evolutionary histories are not in accord with the predictions of the species isolation hypotheses. (Genitalia drawings after Fennah 1967.)

pared with differences in other body characters. We would expect the predictions to be confirmed in some cases simply as a result of lack of uniformity in the rates of evolution of different characters. In addition, several isolated species that have presumably been isolated for especially long periods of time, judging by the fact that they are classified in monotypic genera, also have species — specific genitalia (Table 3.2); there may be a trend toward reduced genitalic complexity in these species, but the sample size is too small to justify confident conclusions. Because there is in general so little confirmation of predictions, I conclude that the data do not support the species isolation hypotheses.

It is still possible that sympatric congeners of the "isolated" spe-

Table 3.2 Species in monotypic genera endemic to islands that have species-specific male genitalia. Nongenitalic characters are also used to recognize species in all cases. Dash indicates that the reference does not give enough information to judge relative complexity of genitalia. (SA = subantarctic islands; G = Galapagos Islands; VI = Virgin Islands.)

Species	Site	Genitalic complexity compared to that of close relatives	References
Coleoptera			
Bountya insularis	SA	Less[a]	Townsend 1971
Orthoptera			
Desmopleura concinna	G	?	Dirsh 1969
Diptera			
Kuschelius dentifer	SA	?	Sublette and Wirth 1980
Pterosis wisei	SA	?	Sublette and Wirth 1980
Hevelius carinatus	SA	?	Sublette and Wirth 1980
Nakataia eisdentifer	SA	?	Sublette and Wirth 1980
Marvella reducta	SA	Less[a]	Sublette and Wirth 1980
Sarothromyiops canus	G	?	Lopes 1978
Galapagomyia inoa	G	More[a]	Lopes 1978
Hemiptera			
Galapagocoris crockeri	G	?	Carvalho and Gagne 1968
Galapagomiris longirostris	G	?	Carvalho and Gagne 1968
Lepidoptera			
Campbellana attenuata	SA	Same[a]	Dugdale 1971
Sorensenata agilitata	SA	Less[a]	Dugdale 1971
Araneae			
Microsa chickeringi	VI	Less[a]	Platnick and Shadab 1977
Darwinneon crypticus	G	Less[a]	Cutler 1971

a. Author makes an explicit statement to this effect.

cies existed in the past but are now extinct, or that zones of sympatry have not been discovered for close relatives (such possibilities can never be completely eliminated), so no given case constitutes definitive proof against the hypotheses. But the number of species and genera that do not conform to the predictions is apparently very large. I estimate that, including the species of other islands, there are hundreds or perhaps thousands, and there are undoubtedly many other mainland cases of isolated, genitalically distinct species restricted to disjunct patches of habitat, for instance, the forest-adapted *Aname* spiders of Australia (Raven 1984).

Parasites

A host is a kind of island, and many host-specific parasites are probably reproductively isolated from even sympatric relatives living on different hosts (Rothschild and Clay 1957; Dritschilo et al. 1975). By examining the host lists of different species that breed exclusively on their hosts, it is possible to determine with reasonable confidence whether given species are isolated from all close relatives.

The use of host lists has three serious pitfalls. The first is that our knowledge of the parasitic fauna of most animal species is incomplete (Ferris 1951; K. C. Emerson, personal communication; R. Price, personal communication), which may cause us to underestimate the numbers of parasitic species to be found on a particular host. The second drawback, accidental cross-contamination (Hopkins 1949; Ferris 1951; Rothschild and Clay 1957; Santana 1976) applies especially to ectoparasites and works in the opposite direction. If, for example, a researcher captures several mammals and carries them in a single sack to a site where they can be searched for parasites, some of the parasites may move to another animal inside the sack. As a result some parasites may be recorded as being less host-specific and thus less isolated from congeners than they really are. Finally, there are natural occurrences of cross-contamination, as when fleas of a prey species jump onto a predator. This kind of contamination must be taken into account because it may result in potential reproductive contact between different species.

I tested the prediction that parasites with no close relatives on the same host should fail to show rapid divergence in genitalic structures with several different groups. Some are discussed below, and others are presented in Table 3.3; see p. 46.

Lice

Lice (Mallophaga and Anoplura) are especially attractive test organisms since they are more host-specific than most other ectoparasites (Webb 1948). In addition, lice are unable to live for more than a short time out of contact with their hosts (Hopkins 1949; Imms 1957), so cross-contamination in nature is probably very rare (Webb 1948; Hopkins 1949; Rothschild and Clay 1957). Another advantage is that a variety of body characters besides genitalia are routinely used in most taxonomic studies.

Emerson and Price (1981) made a current and exhaustive survey of host information for the biting lice (Mallophaga) of mammals. Of a total of 410 species, 236 have never been found on a host species that harbors a congeneric louse species. Nevertheless, the male genitalia of most species, and in some cases also the female genitalia, are generally species-specific (see Hopkins 1949; Werneck 1950; Ferris 1951; Bitsch 1979). This evidence contradicts the predictions of the species isolation hypotheses. It might be possible to argue that our ignorance of mammalian mallophagans is so great that the apparent isolation of many species is illusory. Further information, however, suggests that that is not the case.

The genus *Geomydoecus*, lice of pocket gophers, is especially well studied taxonomically, so data from this group can be used with more confidence. *Geomydoecus* is divided into two subgenera, *Thomomydoecus* and *Geomydoecus*, and in both the genitalia of both males and females have been particularly useful in distinguishing species (Price and Emerson 1972; Price 1974, 1975; Price and Hellenthal 1975a, b, 1976). The two subgenera are kept in the same genus "primarily for convenience, since they are quite different from each other, more so than, say, *Geomydoecus* s. str. is from a number of other trichodectid genera" (R. Price, personal communication). Although species of different subgenera commonly occur together on the same host, two species of the same subgenus rarely occur together. So each subgenus could well be classified as a genus, and species within each subgenus are usually isolated from each other. Furthermore, when more than one species in the subgenus *Geomydoecus* is found on a single host taxon, the different louse species are almost invariably on geographically isolated populations of the host (Price 1972b; Price and Emerson 1972; Hellenthal and Price 1976; also R. Price, personal communication). In sum, a detailed look at an especially well-studied group of lice in which genitalia are generally species-specific confirms the impression that closely related species are often strictly isolated from each other and reinforces the conclusion that the predictions of the species isolation hypotheses are not fulfilled.

Some genera of lice in the Emerson and Price host lists have greater numbers of isolated species than others. Genera with eight or more species, of which a great majority are isolated (numbers in parentheses are of isolated and nonisolated species) are *Neotrichodectes* (8, 2), *Stachiella* (9, 0), *Felicola* (16, 8), *Damalina* (13, 3), *Bovicola* (20, 9), *Tricholipeurus* (13, 7), *Boopia* (12, 2), and *Heterodoxus*

(9, 4). Genera with relatively few isolated species include *Gliricola* (8, 22), *Parafelicola* (1, 7), *Procavicola* (4, 27), and *Geomydoecus* (30, 44). Assuming that the differences among genera are real rather than a product of unequal sampling, the species isolation hypotheses would predict that interspecific genitalic differences should be more common and pronounced in the second group than in the first. Such is not the case (K. C. Emerson, personal communication).

A cursory review of the literature on avian mallophagans revealed several cases that also fail to follow the predictions of the species isolation hypotheses. All of the thirty-two species in the genus *Brüelia* have distinctive male genitalia, but twenty-two of those species, as far as is known, are limited to a single host species or subspecies of Corvidae on which no other *Brüelia* species have been found (Ansari 1956, 1957). Other genera in which male genitalia (and other characters) are generally species-specific and in which at least some species apparently do not share host species with any congeners include the following (numbers in parentheses are of isolated and nonisolated species): *Austromenopon* (9, 1; Fig. 3.3; Price and Clay 1972), *Eomenopon* (12, 3; Price 1966, 1969, 1972a;

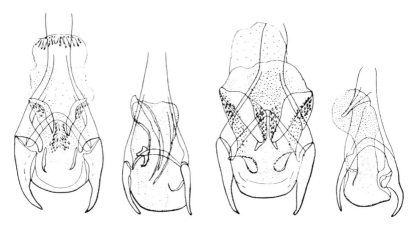

Figure 3.3 The complex, and in some cases asymmetrical, male genitalia of four species of the biting louse genus *Austromenopon*. Each species is found only on a host species that is not known to harbor any other *Austromenopon* species. Despite the apparent isolation of these species from reproductive contact with any close relatives, the male genitalia are complex and species-specific, clearly contradicting the species isolation hypotheses. (From Price and Clay 1972.)

Price and Emerson 1982), *Holomenopon* (7, 9; Price 1971), and *Colpocephalum* (28, 32; Price and Beer 1964, 1965; Price 1976; Price and Emerson 1982). These may be underestimates of the numbers of isolated species; on some relatively well-studied hosts, such as flamingos (Clay 1974), different louse species, with species-specific genitalia, exist on geographically distinct races of the same host species (Clay 1976). There are only a few genera of bird lice in which male genitalia are not good species characters (K. C. Emerson, personal communication). In summary, the evidence from several genera of bird lice contradicts the predictions of the species isolation hypotheses.

Ferris (1951) summarized the host-parasite relations of the sucking lice (Anoplura). His host list shows that, in contrast to the groups just discussed, most anopluran species are not restricted to a single host species but are known from several closely related host species. Each host species, however, usually harbors only a single anopluran species; 369 (80.0 percent) of the 461 host species had only a single anopluran species, and 79.9 percent of the 278 anopluran species (in 40 genera) were not known from any host species that harbored congeneric louse species. Although data are incomplete for this group (Ferris estimated in 1951 that about half of the louse species were known), it appears that many anopluran species have probably evolved in isolation from all congeners. Despite this isolation, male genitalia are useful species characters, along with other characters, in many groups. A partial review of the literature shows this to be true in the following genera (again, numbers in parentheses are of isolated and nonisolated species): *Enderleinellus* (15, 5; Kim 1966), *Neohaematopinus* (8, 0; Johnson 1972a), *Polyplax* (33, 2; Johnson 1972b), and the *hesperomydis* complex of *Hoplopleura* (9, 0; Kim 1965). Thus sucking lice probably represent another failure of the predictions of the species isolation hypotheses.

Parasitic nematodes

Many species of parasitic nematodes mate only within their hosts (Chitwood and Chitwood 1974) and are thus reproductively isolated from species parasitizing other host species. Counts from Yamaguti's (1961) listing of all known species and their hosts suggest that many species may not share the same host species with other congeneric nematodes (672 of 1,406 species in 294 genera for which the host species of all congeners were given). In general, however, nematodes are probably very poorly known, so such counts are not strong evidence. Two groups of nematodes that attack mammals and are better known offer more convincing data.

The nematode genus *Travassosius* consists of two species that parasitize species of beaver, the nearctic *Castor canadensis* and the palearctic *C. fiber*. Bush and Samuel (1978) combined data on the distribution of the nematodes and the probable history of their hosts, and deduced that "the primitive form, *T. americanus*, moved west to eastern Siberia in *C. canadensis* via the Bering bridge, maintained its identity during the evolution of *C. f. pohlei* east of the Ural Mountains . . . and evolved following the evolution of its host to become *T. rufus* in those subspecies of *C. fiber* which occur west of the Ural Mountains in Europe." The derived nematode species is distinguishable only on the basis of its longer spicules (intromittent organs).

The thirteen known species of the intestinal parasite genus *Enterobius* (pinworms) are apparently each limited in nature to a single genus of primates. Although occasional cross-contaminations occur in zoos, no more than one species of pinworm has ever been collected from a given host species in the wild, and the rule of one parasite species to one host genus is thought to hold (see summary of evidence in Brooks and Glen 1982). The worms copulate inside their hosts (Belding 1958). An indication that *Enterobius* species have been reproductively isolated in their hosts is that the parasites' taxonomic relationships closely reflect the relationships of their hosts (Brooks and Glen 1982). Contrary to the species isolation predictions, male genitalia are taxonomically useful throughout *Enterobius*, and Inglis (1961) noted that, although other characters are also used, in most species the spicules "offer the best characters for the delimitation of species"; see Fig. 3.4.

It might be argued that the pinworm evidence is weak, since not all parasite species of some primate groups are known (for example, Cercopithecoidea, Inglis 1961). But those from hosts in Hominoidea *are* relatively well known and clearly show both the pattern of one parasite species – one host genus and species-specific differences in spicule morphology (Inglis 1961).

In summary, data from parasites do not support the predictions of the species isolation hypotheses. The parasite data are particularly damning because they involve both large numbers of species and a high probability that the species have been strictly isolated from

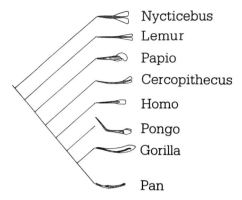

Nycticebus

Lemur

Papio

Cercopithecus

Homo

Pongo

Gorilla

Pan

Figure 3.4 Male intromittent organs (spicules) of *Enterobius* nematodes (pinworms) arranged according to the worms' apparent relationships (Brooks and Glen 1982) and paired with the host genus for each species. Despite their strict reproductive isolation from all congeneric species, these nematodes show rapid and divergent genitalic evolution. (Spicule drawings from Inglis 1961.)

reproductive contact with close relatives. In fact, different evolutionary lineages of some of these parasites have probably been isolated for very long periods of time, yet their genitalia are not simple, and they do show species-specific differences.

It might be argued that the data from taxonomic studies performed for another reason — to distinguish species — cannot be justifiably used to test the hypotheses examined here. But the predictions concern the relative usefulness of genitalic versus other structures in distinguishing species, and these questions are clearly treated in many taxonomic studies. In fact, taxonomists are particularly well trained to judge just such questions accurately. Since the taxonomists compiled their data without knowing that it would be applied to the hypotheses tested here, their data are in effect "blind" and are thus especially useful for testing the hypotheses.

A final possible problem is that not only congeners, but also more distantly related species, might enter into sexual contact frequently enough to cause appreciable selection for genitalic species isolating mechanisms and that my criteria for "isolation" were thus too weak. This seems improbable but is difficult to eliminate in many cases. In some cases, however, it is most unlikely to have been a factor. For instance, *Oryzomys* and *Nesoryzomys* are the only rodents native to the Galapagos, the rhaphidophorid crickets are the

Table 3.3   Genera of parasites (not discussed in the text) in which many species are apparently reproductively isolated from all congeners but have species-specific genitalia. Nongenitalic characters are also used to distinguish species in all groups. Numbers in parentheses are number of apparently isolated species/total number of species in the genus or family.

Genus	Host	References
Nematoda		
Microtetrameres (14/15)	Australian birds	Mawson 1977
Cyclostrongylus[a] (5/7)	Australian marsupials	Beveridge 1982
Metastrongyloidea[b] (5/9)	Australian marsupials	Spratt 1979
Paracooperia (1/4)[c]	Artiodactyls	Gibbons 1978
Johnstonema (1/3)[c]	Australian marsupials	Spratt and Varughese 1975
Dipetalonema (9/16)	Australian marsupials	Spratt and Varughese 1975
Insecta		
Megastrebla, Brachytarsina (6/6)	Bats	Maa 1971
Selenopotes (6/8)	Artiodactyls	Kim and Weisser 1974
Arachnida		
Trouessartia[d] (many/ about 250)	Birds	Santana 1976

a. Grasping organs (bursae) are also species-specific.

b. There is a "substantial degree of definitive host specificity"; only species indigenous to Australia are counted.

c. Spicules (male intromittent organs) of isolated species are longer or more complex than those of other species.

d. The male hysterosomal terminus, which probably contacts the female during copulation (Popp 1967), is also species-specific. Author stated that as many as 1,000 species may exist, but only about 250 were examined; the "majority of the species evidence a 1 host – 1 parasite relationship."

only orthopterans on most of the islands where they are found, *Epichoris* is the only byrrhid on most of the islands they inhabit, and most mammal species are parasitized by only a single species of anopluran louse. In short, this criticism is not important in these cases.

In fact, my criteria for isolation were probably overly strict. One could argue, for instance, that reduced genitalic divergence and complexity should also occur in many parapatric species, since (1) the zone of overlap is often very small with respect to overall species ranges, (2) low mobility makes it unlikely that characters of use only in that zone would consistently spread throughout the species, and (3) parapatry in the present suggests that widespread sympatry in the past was not possible (see White 1978). The species isolation predictions also fail here: in the morabine grasshoppers of Australia, probably the group in which parapatric distribution is best established, male genitalia are one of the two most important features distinguishing species (Key 1981).

The combination of data from parasite species and endemic island species argues clearly against the species isolation hypotheses. Since the predictions tested are derived directly from the logic of the hypotheses, and since some of the groups examined almost certainly fulfill the conditions of strict isolation yet nevertheless have complex, species-specific genitalia, the inescapable conclusion is that neither of the species isolation hypotheses is generally correct. This verdict is particularly impressive in view of the variety of animal groups for which we have data, from rats to roundworms.

Sexual encounters and complexity of spermatophores

Another, independent test of the species isolation hypotheses can be made for some animals that produce spermatophores. In certain species, all of which are arthropods as far as I know, males never encounter females sexually. Instead they scatter their spermatophores in the habitat where females are active, and receptive females later insert spermatophores in their genital openings. The spermatophores of these species are usually spheres containing sperm at the tips of supporting stalks. Apparently the males compete to lay the most spermatophores; there are several accounts of a primitive kind of sperm precedence competition in which males eat the spermatophores of other males that they encounter and then replace them with their own (Weygoldt 1969, on pseudoscorpions; Schaller 1971, on millipeds and collembolans).

In other, related groups the males also place spermatophores on the substrate, but only in the presence of females and often after a more or less elaborate courtship. In some groups, including scorpions, some salamanders, some pseudoscorpions, thysanurans, and some mites, the male maneuvers the female into position over the spermatophore so that she can receive it. The males of some pseudoscorpions have a specialized claw that is species-specific in form

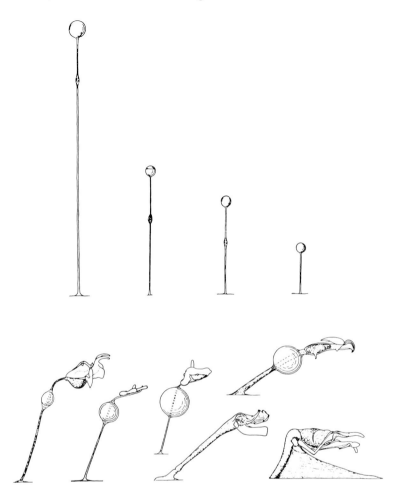

Figure 3.5 Spermatophores of pseudoscorpions. *Above,* males of these species never encounter females sexually but instead scatter spermatophores in the females' general habitat, and females later pick them up. *Below,* spermatophores which are as complex as the genitalia of many insects, made by species in which the male lays a spermatophore only after he encounters and courts a female. The lack of complexity in species without mating and courtship behavior contradicts the predictions of the species isolation hypotheses; the greater complexity in species in which males physically induce females to receive them is in accord with the sexual selection by female choice hypothesis; see Chapter 9. (From Weygoldt 1969.)

to "help" her open her genital aperture (Weygoldt 1969; see Table 11.1). It is usually thought that fertilization without contact between the sexes is ancestral and that contact is derived (Weygoldt 1969, 1970, for pseudoscorpions; Schaller 1971, for hexapods; B. O'Connor, personal communication, for mites). It is also thought that the male behavior of positioning the female directly over the spermatophore is derived from behavior in which the female positions herself (Arnold 1972, for salamanders; Schaller 1971, for apterygote hexapods). The evolutionary relations are probably more complex, however, and lack of direct male-female interaction is probably derived in some groups (Alexander 1964; H. Witte, personal communication, for some parasitogone mites). It is also generally supposed that simple spermatophore morphology is primitive with respect to complex morphology (Weygoldt 1969, 1970, for pseudoscorpions).

Table 3.4 shows a strong correlation between the degree of male-female contact and the structural complexity of the males' spermatophores. Weygoldt (1969) first found this trend in pseudoscorpions, which show both types of mating (Fig. 3.5). He noted that the trend suggests species isolation is not the function of spermatophore complexity (Weygoldt 1970), but offered no explanation as to why it should arise. As noted by both Dobzhansky (1941) and Alexander (manuscript), there is a correlation between the geometric complexity of genitalic structures and their usefulness in species-level taxonomy. The great complexity of the spermatophores in the lower part of Fig. 3.5, for example, leaves little doubt that although the spermatophores of most other congeneric species are not yet known, they will probably show species-specific variation. Such variation is already documented in some groups of pseudoscorpions, uropygids, amblypygids, and scorpions (Alexander 1962a; Weygoldt 1970; 1972; Lourenço 1980; see Fig. 1.8).

The correlation in Table 3.4 is especially important because it directly contradicts the predictions of the species isolation hypotheses. If complex spermatophores serve to isolate species, then spermatophores deposited by males that have had no prior contact with females should be *more* complex rather than less so; no stimuli, other than those associated with the spermatophore itself, are available to help the females of these species identify the species of the male which made the spermatophore, and the danger of interspecific crosses must be especially great (Weygoldt 1970).

Complex spermatophores cannot be explained as adaptations to other environmental factors. In species in which the male contacts a

Table 3.4 Relationship between spermatophore complexity and male-female contact in animals that place spermatophores on the substrate. Contact is counted as occurring if the male contacts the female before depositing a spermatophore, even if the pair has separated at the moment of sperm uptake by the female. Spermatophore complexity is categorized as follows: No — stalk is straight or nearly so and unbranched, sperm mass is spherical; (No) — stalk is not straight, sperm mass is spherical; (Yes) — stalk is straight, sperm mass not spherical but is not structurally complex; Yes — sperm mass is nonspherical and structurally complex, stalk may or may not be straight. Some spermatophores inject their contents into females; others do not but physically stick to females, presumably making sperm uptake more probable.

Organism	Contact between sexes?	Complex spermatophores?	References
Labidostoma cornuta (trombidiformid mite)	No	No	Schuster and Schuster 1969
Tydeus schusteri (trombidiformid mite)	No	No	Schuster and Schuster 1970
Hydryphantes ruber (hydrachenellid mite)	No	No	Mitchell 1958
Polyxenus lagurus (milliped)	No	No	Schaller 1971
Scutigerella sp. (symphylan)	No	No	Kaestner 1968; Schaller 1971
Stylopauropus pedunculatus (pauropod)	No	No	Schaller 1971
Orchesella, 3 spp. (collembolans)	No	No	Mayer 1957; Schaller 1971
Tomocerus, 2 spp. (collembolans)	No	No	Mayer 1957; Schaller 1971
Neobisiidae, 2 spp. (pseudoscorpions)	No[a]	No	Weygoldt 1969, 1970
Garypidae, 3 spp. (pseudoscorpions)	No[a]	No	Weygoldt 1969, 1970
Cheiridiidae, 2 spp. (pseudoscorpions)	No[a]	No	Weygoldt 1969, 1970
Pseudogarypus banksi (pseudoscorpion)	No	No	Weygoldt 1969, 1970
Larca sp. (pseudoscorpion)	No[a]	No	Weygoldt 1969, 1970
Oribatid mites, 17 genera, 22 spp.	No[b]	(No)[c]	Taberly 1957; Woodring and Cook 1962
Trombicula, 3 spp. (chiggers)	No	(No)	Lipovsky et al. 1957
Hannemania sp. (chigger)	No	(No)	Lipovsky et al. 1957
Campodea remyi (dipluran)	No	(No)	Engleman 1970; Schaller 1971
Anystis baccarum (trombidiformid mite)	No	(No)	Schuster and Schuster 1966
Sphaerolophus sp. (erythraeid mite)	No	(No)	Witte 1977
Diptacus gigantorhynchus (eriophyid mite)	No	(No)	Oldfield et al. 1970
Novophytoptus sp. (eriophyid mite)	No	(No)	Oldfield et al. 1970

Species			Reference
Aceria sheldoni (eriophyid mite)	(Yes)	No	Sternlicht and Goldenberg 1971
Eriophyes, 3 spp.	(Yes)	No	Oldfield et al. 1970; Sternlicht and Griffiths 1974
Aculus cornutus (eriophyid mite)	(Yes)	No	Oldfield et al. 1970
Phyllocoptruta oleivora (eriophyid mite)	(Yes)	No	Oldfield et al. 1970
Chthoniinae, 2 spp. (pseudoscorpions)	(Yes)[d]	No	Weygoldt 1969, 1970
Bdellidae (mites)	Yes	No	Wallace and Mahon 1972, 1976; Alberti 1974; M. M. H. Wallace, pers. comm.
Halacaridae (mites)	Yes	No	Pahnke 1974
Pseudotriton sp. (salamander)	Yes	No[c]	Organ and Lowenthal 1963; Arnold 1972
Notophthalmus viridescens (salamander)	Yes	No	Jordan 1891; Arnold 1972
Arrenurus globator (hydrachenellid mite)	Yes	No	Bottger 1962; Schaller 1971
Dicyrtomina minuta (collembolan)	Yes[e]	No	Mayer 1957; Schaller 1971
Sminthurides, 2 spp. (collembolans)	Yes	No	Mayer 1957; Wallace and Mackarras 1970
Podura sp. (collembolan)	Yes	No	Englemann 1970; Schaller 1971
Serianus sp. (pseudoscorpion)	Yes[f]	(No)	Weygoldt 1969
Geophilus longicornis (centipede)	Yes[f]	(No)	Klingel 1959
Olpium sp. (pseudoscorpion)	Yes[a]	(No)	Weygoldt 1969
Lepisma saccharina (silverfish)	Yes[f]	(Yes)[g]	Sturm 1956
Ambystoma, 9 spp. (salamanders)	Yes	(Yes)	Mohr 1931; Noble and Brady 1933; Shoop 1960; Anderson 1961; Uzzell 1969; Spotila and Blumer 1970; Arnold 1972; Garton 1972
Erythraeid mites, 3 genera, 4 spp.	Yes	(Yes)	Putman 1966; Witte 1975
Saxidromus delamarei (saxidromid mite)	Yes	(Yes)	Coineau 1976
Scolopendra cingulata (centipede)	Yes	(Yes)	Klingel 1957
Lithobius forficulatus (centipede)	Yes	(Yes)	Schaller 1971
Scutiger coleoptrata (centipede)	Yes	(Yes)	Klingel 1956
Thereupoda decipiens cavernicola (centipede)	Yes	(Yes)	Klingel 1962
Plethodon, 3 spp. (salamanders)	Yes	(Yes)	Organ and Lowenthal 1963; Arnold 1972
Scorpions, 5 families, 16 genera, 26 spp.	Yes	Yes[g]	Kaestner 1968; Francke 1979; Lourenço 1980

Table 3.4 (continued)

Organism	Contact between sexes?	Complex spermatophores?	References
Telyphonus caudatus (uropygid)	Yes	Yes	Schaller 1971
Mastigoproctus, 2 spp. (uropygids)	Yes	Yes[h]	Weygoldt 1972
Trithyreus sturmi (schizomid)	Yes	Yes[g]	Schaller 1971
Damon variegatus (amblypygid)	Yes	Yes[i]	Alexander 1962b
Tarentula marginemaculata (amblypygid)	Yes	Yes[i]	Weygoldt et al. 1972
Charinus brasilianus (amblypygid)	Yes	Yes	Weygoldt 1972
Sarax sarawakensis (amblypygid)	Yes	Yes[i,j]	Schaller 1971
Ademetrus, 2 spp. (amblypygids)	Yes	Yes[i,k]	Alexander 1962a; Weygoldt 1972
Dactylochelifer latreilli (pseudoscorpion)	Yes	Yes	Weygoldt 1969
Dinocheirus tumidus (pseudoscorpion)	Yes	Yes	Weygoldt 1969
Chelifer cancroides (pseudoscorpion)	Yes	Yes[g]	Weygoldt 1969
Atemnus politus (pseudoscorpion)	Yes	Yes[g]	Weygoldt 1969
Chernes cimicoides (pseudoscorpion)	Yes	Yes[g]	Weygoldt 1969
Withius subruber (pseudoscorpion)	Yes	Yes[g]	Weygoldt 1969
Proteus anguinus (salamander)	Yes	Yes	Briegleb 1961
Allothrobium spp. (mites)	Yes[?l]	Yes[?l]	Moss 1960; Evans et al. 1961; Mather and LeRoux 1970; Schaller 1971

a. Male is not in contact with female when she picks up spermatophore, but either male or spermatophore apparently emits attractive odor.

b. In some species rate of spermatophore deposition increases when females are present with males.

c. Spermatophore is sticky.

d. Weygoldt (1969) speculated that spinelike processes around spermatophore serve to protect the sperm when female investigates spermatophore.

e. Male builds a "fence" of spermatophores around female but apparently does not otherwise induce her to inseminate herself.

f. Male builds web, is not in contact with female as she picks up spermatophore.

g. Spermatophore injects sperm into female.

h. Male presses spermatophore into female opening with his palps.

i. Female "rips" sperm package from spermatophore.

j. "Sperm plates" and "sclerites" in spermatophore are mentioned, but complete description is not given.

k. Male evidently opens sperm package with chelicerae before female takes it into her gonopore.

l. Moss (1960) indicates lack of male-female contact and simple spermatophore morphology in *A. lerouxi.*

female and then makes a spermatophore immediately before the female takes it up, the spermatophore must survive intact for only a brief period of time before being received, and other animals are not likely to damage it. On the other hand, spermatophores deposited by males in the absence of nearby females are subject to selection for resistance to the vicissitudes of the environment. Thus any adaptations to resist desiccation and to protect against other animals should be found in the spermatophores deposited by males that do not contact females; however, most of the spermatophores in this category have only a thin covering (Schaller 1971, for mites; Weygoldt 1969, for pseudoscorpions). The most likely explanation for spermatophore complexity, as I will show in Chapter 9, is that the complex spermatophore structures are "internal courtship" devices designed to encourage partially receptive females (probably via genital stimulation) to receive them and utilize the male's sperm to fertilize their eggs; such devices would be superfluous in species without male-female contact since only receptive females ever bring their genitalia into contact with spermatophores.

Left unanswered is the question of how species isolation occurs in most of the groups that lack male-female contact. I found no reports of observations on whether females of these groups can distinguish spermatophores of their own species from those of others prior to picking them up with their genitalia. The simple spermatophores of some oribatid mites and some pseudoscorpions differ slightly between species (Taberly 1957; Evans et al. 1961; Weygoldt 1970). Those of a number of collembolans, on the other hand, are quite uniform (Mayer 1957). It seems likely that females use as-yet-undetected chemical signals as species-specific cues; females of some mites and perhaps some pseudoscorpions appear to use chemical cues to find spermatophores (Putman 1966; Witte 1975; Weygoldt 1969, 1970).

4 Tests of the other hypotheses and a summary

Pleiotropism

Mayr's (1963) hypothesis has two parts. First, the differences in genitalic design of closely related species are thought to be selectively neutral — that is, sperm can be transferred equally well by all of them. Second, rapid, divergent evolutionary change occurs in genitalia because their form is determined at least in part by genes that also influence other body characters. Selection on these other characters thus results in gene changes that incidentally cause changes in the genitalia.

There are theoretical reasons to doubt the claim of selective neutrality, and there are data that contradict the idea of pleiotropism. The question of selective neutrality at the level of molecules is still hotly debated in evolutionary biology, but even strong proponents of neutralism (Kimura 1979) specifically avoid claiming selective neutrality for higher-level structures such as tissues and organs. And with good reason, because the analysis of organic structures in terms of their functional properties has been, and continues to be, an extremely fruitful branch of biology. In some cases even bizarre and seemingly meaningless details of structure have been found to have functional significance. An instructive example occurs in fiddler crabs, *Uca*. The males have one greatly enlarged claw (see Fig. 4.1) which they use in aggressive interactions with conspecific males. In many species the claws have complex and species-specific patterns of ridges and pimples, minor details in surface sculpture that would seem especially likely to be neutral characters. Detailed studies of male agonistic behavior (Crane 1975) have shown that in

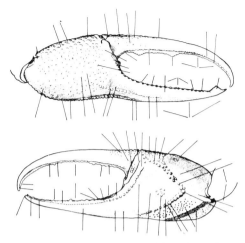

Figure 4.1 The outer and inner surfaces of the claw of a male fiddler crab
(Uca) with a cautionary tale for those who doubt the importance of natural
selection. The surface carries an array of bumps, ridges, and depressions
that do not appear to have any functional importance. Intensive study,
however, has shown that many of these details (indicated by lines) serve as
signaling structures in male-male aggressive interactions. Studies of other
apparently useless animal structures will undoubtedly reveal many more
cases of functional significance. (After Crane 1975.)

fact these bumps are part of a complex tactile signaling system for
resolving combats. Equally detailed studies that could test functional explanations of genitalic characters have not yet been performed for most animal species. Thus it is premature to reject the
possibility of functionality, which been demonstrated in so many
other kinds of organs.

Neutrality also seems unlikely for another reason. Natural selection focuses on differential reproduction, favoring those organisms
that reproduce best. For instance, a male cat that is a superior run-
ner, hunter, and fighter but that fails to induce ovulation when he
mates with females is an evolutionary failure. It is risky at best to
assume selective neutrality for the male organs that are directly
involved with both the placement of the male's gametes in the
female and, probably in many cases (see Chapter 7), with triggering
the female reproductive processes that determines whether those
sperm will fertilize her eggs.

In addition to these theoretical arguments, factual evidence
strongly contradicts Mayr's pleiotropism hypothesis. This might
seem surprising, given our ignorance of the genetic determination

of genitalic morphology. The only studies I know of are incomplete: Gordon and Rosen (1951) showed that a small number of polygenes controlled the development of the "claw" on the gonopodium of the poeciliid fish *Xiphophorus* and *Platypoecilus;* Turner and colleagues (1961) showed that the presence of a genitalic spine in the butterfly *Papilio dardanus* is controlled in the main by a single gene; and Peck (1983) found no intermediates in genitalic morphology when he crossed different species of the leiodid beetle *Ptomaphagus.* None of these studies mentioned pleiotropic effects.

There are, however, data that relate to the chances that genitalia, and not other organs, should be incidentally affected pleiotropically. Perhaps the most obvious question is why genitalia are so consistent in evolving rapidly and divergently, when pleiotropic effects are thought to be more or less randomly distributed among body characters. A common answer is that genitalia are especially complex structures and are thus probably influenced by especially large numbers of loci, so more pleiotropic effects are expected. But this answer, I think, only dodges the issue, leaving unexplained the question of why genitalia are complex in the first place. Why, to take a somewhat similar structure as an example, is the anus so seldom a useful species character? The selective neutrality of the details of its structure is, if anything, more certain than that of the genitalia, yet anal structure is seldom useful for distinguishing species. Pleiotropism thus does not adequately explain the overwhelming tendency, in animals with internal fertilization, for genitalia rather than other organs to be especially useful characters to distinguish species.

Another widespread trend is also unexplained by pleiotropism. The genitalia of animals with external fertilization (for example, echinoderms, most fish, and most polychaete worms; see Table 1.4), are not elaborate and do not have species-specific forms. One would confidently expect the details of genitalic structure in these groups to be very close to selectively neutral, but the expected pleiotropic effects are absent.

Finally, and most important, in a number of animal groups sperm is transferred by secondary genitalia, such as legs or chelicerae, or by spermatophores, rather than by the male's primary genitalia. In group after group of this type, the primary genitalia are not useful species characters, while the secondary genitalia or spermatophores are (Tables 1.2 and 1.3; see also Scudder 1971). That the supposed pleiotropic effects should fail to appear in just those primary genitalia not involved in copulation and should be so consistently present in just those structures used in copulation defies explanation in terms of pleiotropism. The theory supposes that there are only chance associations between allele changes and their pleiotropic effects on genitalia, but the association between rapid and divergent evolution and direct contact with mates is so strong that it discredits the theory.

Arnold's modification

To avoid some of these criticisms, Arnold (1973) modified the pleiotropism hypothesis. He abandoned the selective neutrality idea and supposed that the genitalic anatomy of each sex is adjusted to that of the other. This idea, however, results in logical problems for his argument. When genitalic form is linked pleiotropically with selectively important nongenitalic characters, then disadvantages resulting from changes in genitalic form will counterbalance the positive pleiotropic effects; thus only changes with relatively dramatic pleiotropic advantages will be favored; since such large benefits are unlikely, consistent rapid evolutionary change, as is typical of genitalia, is unlikely. In addition, the compensatory genitalic changes that he thought should occur would often be accompanied by changes in these other characters. By Arnold's own reasoning, this would probably be deleterious. Thus the compensatory changes should also occur less often and more slowly than Arnold supposed. There is also widespread evidence that male and female structures often fail to evolve in parallel as required by Arnold's argument. As I discussed in Chapter 3, female structures are in many cases more or less invariable while male structures vary dramatically, and occasionally the reverse is true.

Mechanical conflict of interest

The mechanical conflict of interest idea has been only briefly outlined by Alexander (manuscript), Lloyd (1979), and Wing (1982); it has never been developed in detail. It is radically different from the other hypotheses and fits well with some of the general trends in genitalic evolution that will be discussed in the next chapter, so it merits careful consideration. The idea is that a male's behavior sometimes favors his own reproductive interests at the expense of those of the female. An example could be the walking stick *Callinda bicuspis,* since males copulate with females for many hours and

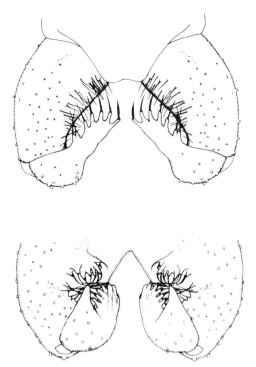

Figure 4.2 The grotesque grasping male genitalia of two species of *Tricho-tanypus* gnats (Chironomidae). These apparently "brutal" male structures symbolize the hypothesis that rapid genitalic divergence is the result of an evolutionary race between males and females to mechanically control events associated with copulation and fertilization. (After Wirth and Sub-lette 1970.)

thereby apparently prevent the females from ovipositing. The long couplings probably function to defend the female from other males but, in contrast to unaccompanied females, which lay about one egg per hour, copulating females do not lay any eggs, and when released they lay a series of eggs unusually rapidly (up to fifteen per hour: D. Windsor, personal communication). Females and males are thought to be in an evolutionary race with each other to develop morphologies (see Fig. 4.2) that permit control over processes associated with copulation such as immobilization of the female and use of sperm for fertilization. I will show here that there are both theoretical and practical reasons to doubt that this is a general explanation for rapid and divergent genitalic evolution.

On the theoretical side, it seems doubtful that mechanical conflict of interest would consistently lead to rapid and divergent evolution. It would not always be adaptive for females to evolve countermeasures to male manipulations, and thus one would not expect unending evolutionary races between males and females to occur. Take, for example, the evolution of devices on the genitalia of male damselflies that remove sperm deposited in previous matings from the female storage organs, or spermathecae (Waage 1979; see Fig. 2.3). Once the males of a species have acquired the ability to clean out spermathecae, the mechanical conflict of interest hypothesis predicts that females would evolve more complex spermathecae so that the male genitalia could not reach all the sperm. But what would happen to a female with a variant spermatheca that was more difficult to clean? Those males that were well adapted to clean typical spermathecae would be unable to clean out this female's spermatheca as completely, putting other males that were less well adapted cleaners at a smaller disadvantage. The variant female's eggs would, on the average, be fertilized more often by males with poorer cleaning abilities (that is, poorer ability to clean out "typical" females) than would the eggs of other females. If males varied genetically in their ability to clean out spermathecae, the variant female's sons would, on average, be *poorer* cleaners than the sons of females with the "typical" spermathecal form, and selection would consequently act against the variant female.

A qualification must be made to this argument. If the variant female gained more in fitness by preventing male manipulation than she lost through having atypical sons, female countermeasures would be favored. This kind of situation could arise as a result either of particularly damaging male manipulations (such as temporarily removing the female from oviposition sites), or of reduced genetic variance in males' abilities to manipulate females, so that the sons of atypical males would not be inferior. This might result in a halting, perhaps relatively slow, evolutionary sequence in which female modifications originated only when genetic variance in male characters had reached, or nearly reached, zero, with subsequent male adjustments arising only slowly through stochastic processes. There is to my knowledge no data on genetic variance in such male genitalic characters or measurements of the reproductive costs to females of male manipulations, so the probability of this type of situation arising cannot be judged on these grounds.

Presumably, given cases could fall on either side of the balance between male and female advantage. The ability to clean out spermathecae, for instance, would seem likely to affect male success more strongly than female success, making female modifications disadvantageous. Thus the conflict of interest hypothesis would not be expected to apply in damselflies. This prediction is apparently correct: in several species of damselflies the shapes of spermathecae are relatively simple, the duct leading to the spermatheca is short and simple, and there are only rather minor variations in spermathecal shape within many genera (Waage 1979, in press) (in contrast see the spermathecae in Fig. 6.8). If the conflict of interest hypothesis is indeed not applicable to these damselflies, then the generally species-specific morphology of male penes in some odonate groups (Kennedy 1919, 1920) is not easily explained by the conflict of interest theory.

In other cases females might gain more by avoiding male manipulation. The ability of some males to sequester females from possible contact with other males before the female oviposits could conceivably be so damaging to a female that her ability to escape would more than compensate for future reductions in her sons' reproductive ability. Again, no data are available on relative selective advantages, making it difficult to evaluate the hypothesis. In sum, conditions sometimes could favor females that escape males' manipulations, but at other times they would act against such escapes, so the generality of the hypothesis is questionable. In damselflies predictions are not congruent with the data, but in most cases there are simply no data available to make the tests.

One further point is that in many (most?) species a female can reject a male before genital coupling by moving away, bending her abdomen out of reach, failing to open her genital aperture, or some other such maneuver. Thus many male-female conflicts of interest are probably resolved behaviorally, without the animals' genitalia becoming directly involved. Female ability to resolve such conflicts early rather than late in male-female interactions would be favored (Alexander 1962, 1964; McGill 1977; Daly 1978). In many such cases morphological genitalic modifications to overcome those of the opposite sex would appear to be superfluous (see also Fig. 6.3).

In addition to these theoretical considerations, several types of data contradict this hypothesis. One important pattern has already been discussed in Chapter 3. The mechanical conflict of interest idea supposes that complementary changes occur in male and female genitalia, but in a variety of groups female morphology is uniform while that of males is diverse.

Another important point is that, as far as I know, not a single example of a female "anticlasper" device is known. In arthropods, for example, a simple, erectable ridge or spine in the region where males grasp females would provide a defense against clasping, yet such structures are either rare or do not exist. Instead, some females (for example, various odonates, Corbet 1962; cicindellid beetles, Freitag 1974; and some bees, Toro and de la Hoz 1976) have species-specific *indentations* to receive the male organ in the area that is grasped (see Figs. 4.3 and 11.2). While it must be admitted that the

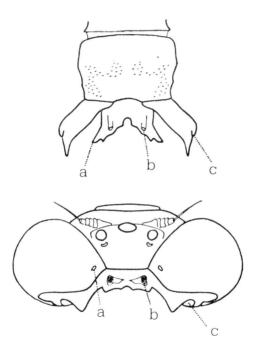

Figure 4.3   Above, the male abdominal appendages and, *below,* the sockets into which they fit on the head of the female *Epigomphus quadracies* dragonflies (parts that interlock are labeled with the same letters). The male seizes the female with his abdominal appendages, which in many groups have species-specific forms. Contrary to the prediction of the mechanical conflict of interest hypothesis, the females generally have indentations that *help* the male maintain his grip rather than make it more difficult for him. (Modified from Corbet 1962.)

detailed workings of many genitalic structures are not understood, I think it is reasonable to read the female modifications just mentioned as not having the generalized anticlasper function that one would expect according to the mechanical conflict of interest hypothesis.

A similar objection stems from examination of some other male genitalic structures, which show a notable lack of features that might be conflictive. Among many possible examples are the complex structures on the pseudoscorpion spermatophores shown in Fig. 3.5, the species-specific brushlike structure on the genitalia of bumblebees *(Bombus)* and *(Cycloneda* ladybird beetles (see Fig. 10.2), which remain outside the female during copulation (Richards 1927a; O. Trejos and W. Eberhard, unpublished), and the special pattern of discs on the male octopus arm (hectocotylus), which probes briefly into the female's mantle cavity and then transfers his grenadelike explosive spermatophore (Wells and Wells 1977). How could any of these possibly act to contravene the female's reproductive interests? Again, the detailed workings of these and other species-specific genitalic structures are still unknown, but their morphologies and what is known of the behavior associated with them give no hint of any conflictive functions.

Finally, in a few cases species-specific male genitalic structures that were formerly called claspers and thought to function in ways that could have resulted in mechanical conflict of interest have turned out to have other functions. For example, the male "claspers" of several butterflies have species-specific patterns of teeth and spines, but rather than holding the female tightly during copulation they rub back and forth against her abdomen (Lorkovic 1952; Scott 1978; Platt 1978) (see Fig. 10.1). A similar situation may occur in nematodes. The male bursa (an extension of the posterior end) is generally thought to be a holding organ and is often useful in distinguishing species (de Coninck 1965). But careful observation of copulation in *Nematospiroides dubius* showed "a localized muscular contraction at the base of the bursa, which pulls the . . . female close to the male. The rays of the bursa [which are particularly useful taxonomically] clearly transmit the contractions . . . but . . . do not seem to be the major holding organs" (Croll and Wright 1976, p. 1469).

It is important to note that a male could also stimulate a female in ways that induced a resolution of male-female conflict in his favor. This in fact is part of the argument that will be introduced in the next

Table 4.1 Summary of evidence against the various hypotheses proposed to explain rapid and divergent genitalic evolution. Parentheses indicate that the data show a trend that is not explained by the hypothesis but that does not directly contradict it.

Lock and Key

Selective context is probably relatively uncommon.

Selection favors earlier species discrimination by females.

Female structures often do not evolve in step with those of males.

Female structures of many groups cannot exclude males of other species.

Rapid genitalic divergence occurs in species isolated from all close relatives.

Correlation exists between spermatophore complexity and male-female contact.

Genitalic Recognition

Selective context is probably relatively uncommon.

Selection favors earlier species discrimination by females.

Rapid genitalic divergence occurs in species isolated from all close relatives.

Correlation exists between spermatophore complexity and male-female contact.

Pleiotropism (Mayr 1963)

Selective neutrality in genitalia is not likely.

Reason why pleiotropism acts on genitalia rather than other organs is not explained.

Rapid divergent evolution does not occur in primary genitalia when other structures transfer sperm.

Rapid divergent genitalic evolution does not occur in species with external fertilization.

(Correlation exists between spermatophore complexity and male-female contact.)

Pleiotropism (Arnold 1973)

Compensatory changes are not as likely as proposed.

Female structures often cannot exclude incorrect males, and compensatory adjustments are thus not clearly necessary.

Female structures often do not evolve in step with those of males.

(Correlation exists between spermatophore complexity and male-female contact.)

Mechanical Conflict of Interest

Sexual selection often favors females not overcoming male manipulations.

Female structures often do not evolve in step with those of males.

Obvious anticlasper devices are lacking in females.

Some male genitalia are nonmanipulative.

chapter, and in that respect the hypothesis of sexual selection by female choice also represents a male-female conflict of interest. But the Alexander and Lloyd hypothesis refers solely to mechanical interaction between male and female genitalia (see Chapter 5 for the relationship between their ideas and sexual selection by female choice).

Summary

Table 4.1 summarizes the evidence presented up to this point. All four previous hypotheses are clearly unable to explain sizable amounts of evidence, and I think it is reasonable to conclude that none is truly general in scope. It is of course possible to rescue any of these hypotheses by claiming that it is true for certain cases that so far show no contradictory evidence. I do not claim that none of them is true for any case, and in the end, of course, it is impossible to eliminate any hypothesis that is progressively weakened in this way to accommodate conflicting evidence (Popper 1973). But there is a central evolutionary phenomenon of sweeping generality: male structures that are specialized to contact females in sexual contexts have an extraordinary propensity to evolve rapidly and divergently. The generality of the phenomenon suggests that there is an equally general reason for its occurrence. The next chapter presents a new theory that may prove equal to the task.

5 Sexual selection by female choice

The emphasis up to this point has been on examining other people's ideas about why genitalia evolve rapidly and divergently. The accumulation of evidence indicates that none of these hypotheses is a good general explanation. I have developed an alternative hypothesis — sexual selection by female choice — which fits the available data better.

The function of copulation

The hypothesis of sexual selection by female choice is based on a reinterpretation of the function of copulation. I will begin by giving the evidence that led to this change in focus. If asked about the function of copulation (at least in nonhumans!), most people would unhesitatingly respond that it serves to transfer gametes from one individual to another and thus to bring about fertilization. But some animals show "perverse" behavior that does not fit this explanation. For instance in some spiders that have secondary male genitalia (pedipalps) that must be loaded with sperm from the male gonopore in order to inseminate females, the males always copulate once before charging their palps, then charge them with sperm and copulate again (van Helsdingen 1965; see Fig. 5.1). Other spider and milliped males also perform preliminary copulations before priming their secondary genitalia with sperm (Kullmann 1964; Haacker and Fachs 1970). Mice and rats mount and intromit several times before finally ejaculating (McGill 1977) (see Chapter 10 for other examples). These behaviors suggest that copulation may have some

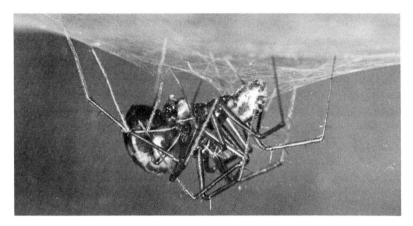

Figure 5.1 The "perverse" copulation of a pair of *Lepthyphantes leprosus* spiders. The male can inseminate the female only after loading his pedipalps with sperm emitted from the primary genital opening on his abdomen. However, the male always "copulates" with the female a number of times before charging his palps with sperm, suggesting that sperm transfer is not the only function of copulation. (From van Helsdingen 1965.)

other function in addition to insemination. I argue here that the other function is that of inducing females to receive and use the sperm or, in a broad sense, of courtship.

A very widespread pattern in male versus female genitalic structure supports this interpretation. The pattern is so nearly universal, in fact, that its importance has been overlooked. In nearly all species of multicellular animals with internal fertilization, it is the male (the sex that produces smaller gametes) that has the intromittent organ or makes the gamete package (spermatophore), which is introduced into the partner. In addition, where there are specialized genitalic organs or secretions that hold the male and female together during or just prior to copulation, it is the male that possesses them; examples include the claspers of many insects (Scudder 1971); holdfasts and suckers in nematodes (Hope 1974); and slime and adhesive secretions in mites (Böttger 1965; Schaller 1971), oligochaetes (Stephenson 1930), and molluscs (Fretter and Graham 1964). This pattern holds even in hermaphrodites such as oligochaetes (Stephenson 1930), molluscs (Fretter and Graham 1964), and chaetognaths (Ghirardelli 1968); for instance, the setae and spines that are thought to hold earthworms together as they copulate are associated with the male rather than the female gonopore. The general pattern that males are more aggressive than females in the sexual activities preceding mating (discussed at length by Darwin 1871 and confirmed many times since: Richards 1927b; Daly and Wilson 1978) seems to extend to the morphology of the animals' genitalia.

Internal fertilization has probably evolved many different times, and in each case the pattern has been the same. I have not attempted to make a careful study of the probable number of origins of internal fertilization, and uncertainties regarding the derivations of some groups would make the results only tentative in any case. Data from three groups with mostly external fertilization but in which internal fertilization has arisen several times independently show that the male sex usually is the one to evolve intromittent organs. Of fifteen families or superfamilies of fish that have species with internal fertilization (Breder and Rosen 1966), in all but the syngnathids (see below) fertilization occurs in the female's body. Intromittent organs occur in four families of polychaete worms, in all cases in the males, and in five other families males make and transfer spermatophores (Schroeder and Hermans 1975). Internal fertilization has arisen independently at least seven times in molluscs (Fretter and Graham 1964; Purchon 1968), and although many species are hermaphrodites, in all cases the intromittent organs transfer sperm rather than eggs.

The fact that holding and intromitting genitalia are restricted to males calls into question the widespread, intuitively reasonable, and often unconscious assumption that once a pair of animals has gone through the process of mate-seeking and courtship and has begun to copulate, both male and female are attempting to bring about the same result — zygote formation. If this assumption were true, then at least some animal species among the million or more that practice internal fertilization should have females rather than males with intromittent organs and/or claspers. The lack of such organs in females suggests that it may be incorrect to assume that males' and females' interests are identical.

Why do females so consistently lack intromittent organs? Why do they consistently receive rather than donate gametes? It cannot be because eggs are more difficult to transfer than sperm, since females of many species do transfer eggs when they oviposit. And even if difficulty of transfer were a problem, the female could sometimes provide the tube for sperm transfer. As with many other evolutionary problems, the probable answer to this question can be

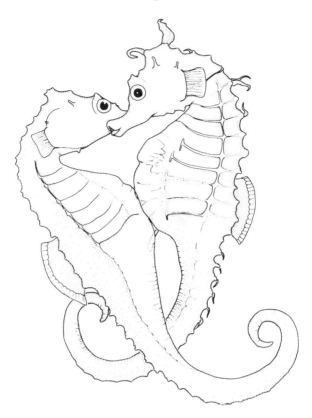

Figure 5.2 A pair of seahorses copulating and thereby breaking the rules. The female, *left,* has inserted her intromittent organ into the male's pouch, where she deposits her eggs. The male will fertilize the eggs there and brood the resulting zygotes. The correlation here between the reversal of sex roles in parental investment and the reversal of male and female genitalic morphology suggests a courtship function for intromittent genitalia.

found by using the comparative method and examining exceptions to the rule.

In this case there is only one certain exception, in syngnathid fish — seahorses and pipefish (see section on topology of female organs in the next chapter for other possible cases). In these fish the female has an intromittent organ which she uses to introduce her eggs into a groove or cavity in the male, where they are fertilized (Fig. 5.2). These fish also reverse sex roles in other respects. The male broods the fertilized eggs, and in some species he has a "placenta" through which the embryo receives nourishment (Greenwood 1975; Ridley 1978). The eggs are large and yolky (Ridley 1978), so it is not certain whether the male invests more in the young than the female does, but this seems probable in at least some species in which the offspring are quite large when they finally emerge from the male's pouch (Bellomy 1969; Greenwood 1975). The coincidence of the apparent reversal of male and female investment in offspring and the reversal in genitalic morphology is striking. It suggests that the usually smaller male investment in offspring may be responsible for the pattern of males having the holding and/or intromittent genitalia.* This argument follows the now widely accepted reasoning that males of most species are less selective and coy in courtship because they make smaller investments in offspring (Darwin 1871; Trivers 1972). A consequence of this difference in investment is that male reproduction is generally more limited by sexual access to mates than is that of females (a male can father the broods of many females; a female cannot usually increase her reproduction by copulating with additional males). Thus the smaller-investment sex is more often obliged to compete with other members of the same sex for mates, giving rise to sexual selection which favors those individuals that win out in this competition. I am suggesting that competition between males for fertilization is responsible for males having intromittent organs.

Another body of evidence is related to this same point. Copulation (genitalic coupling of male and female) is commonly assumed to be equivalent to (or always result in) insemination (deposition of sperm in the female); and insemination is assumed in turn to necessarily result in fertilization. In reality, however, these three processes are usually uncoupled in both time and space. Copulation can, and does, occur naturally without resulting in insemination, and insemination can, and does, occur naturally without resulting in fertilization (these points are documented in Chapter 7). As shown in Tables 1.1–1.3, males of species with internal fertilization rarely if ever deposit their sperm directly onto the eggs. There may be several adaptive reasons for this uncoupling, and probably the advantage to the female of not having the eggs fertilized until she can lay them in a protected or otherwise advantageous site is important. Whatever the reasons for delayed fertilization, it has the important consequence that a female can copulate with a male without all of her

* The probable reason why females of other species which have paternal care and internal fertilization nevertheless lack intromittent organs is discussed in Chapter 10.

eggs necessarily being fertilized by his sperm. Copulation is not the end of the fertilization process, and since the subsequent events occur within the female's body, she rather than the male can influence them directly.

The question of why genitalia evolve rapidly and divergently is obviously linked to how genitalia function. The two ideas just presented — that males use their genitalia competitively to increase their chances of obtaining fertilizations, and that copulation does not always (or even perhaps usually) lead to fertilization of all the female's eggs — require a change in thinking about how genitalia function in the processes of copulation, insemination, and fertilization. Mating is evidently not always a completely cooperative process to insure fertilization; males sometimes attempt to inseminate incompletely receptive females; and female behavior during and after copulation can determine whether a male that has copulated with her will sire any of her offspring.

The female choice hypothesis

The female choice hypothesis, in capsule form, is that females discriminate among males of their own species on the basis of the males' genitalia, and that males with favored genitalic morphologies sire more offspring than others. The consequence, following the arguments of Darwin (1871) and Fisher (1958), is that runaway sexual selection by female choice on male genitalia results in both rapid and divergent evolution. In short, I am suggesting that male genitalia function as "internal courtship" devices and that their use is properly considered as an extension of the male's courtship of the female.

How could female discrimination based on male genitalia arise? The genitalia of some males of a species may be more effective in entering the female or in holding the male in place so that he introduces more sperm into an advantageous position inside the female (where the sperm will be more likely to fertilize the eggs). Or stimuli from the genitalia of some males may be more effective in eliciting essential female reproductive processes, so that copulation is not terminated prematurely, sperm is transported to storage and/or fertilization sites, ovulation occurs, the eggs mature, stored sperm is nourished, the fetus is implanted, or further attempts at copulation are resisted (see Chapter 7). Those males that are mechanically superior or are better stimulators are favored; in consequence, those females that preferentially allow such males to fertilize their eggs are also favored, because their sons will be especially good at siring offspring. A female could discriminate among males' genitalia on the basis of their ability to fit mechanically into hers, or through other sensations occurring in her genital region. The ability to discriminate among males using mechanical or stimulatory genitalic cues could thus spread genetically through the females of a population. Once such female discrimination was established, selection would favor any male that was better able to meet the females' criteria (by squeezing her harder, touching her over a wider area, rubbing her more often, and so on) even though his genitalia were no better at delivering sperm than those of other males: the ability to "convince" the female would become in itself an important determinant of male fitness.

Females that discriminated in favor of more stimulating males would be favored because their sons would be more likely to be effective stimulators and would thus be likely to sire more offspring. It is important to keep in mind that (1) the male need not be superior in any respect other than the ability to stimulate or fit with the female in order for selection to operate on both sexes to increase the importance of these cues in determining male reproductive success; and (2) this does not require a complex female "aesthetic" sense such as that apparently envisioned by Darwin (1871), but only an ability to respond differentially to sometimes quite simple stimuli.

A key aspect of this model of genitalic evolution (and of runaway evolution by female choice in general) is the arbitrary nature of the cues used by females to discriminate among males. Once a male stimulus elicits responses in females that improve the male's reproduction, the stimulus itself (and the ability to increase its effectiveness) becomes an advantage per se, regardless of any possible association with other aspects of male quality. Even if a particular stimulus is originally advantageous due to association with male superiority in some other respect (for example via pleiotropy), once the stimulus is successful (that is, used by females in determining whether or not fertilization will occur), its stimulus value will become an advantage in itself.

As an illustration, take a population in which females are discriminating among males on the basis of a particular stimulus (call it *A*). If any male can deliver a "superoptimal" stimulus (say *A* plus *B*) that increases the female's receptiveness, he will be a superior reproducer even if he is only average in all other characteristics, as is likely. Those females best able to discriminate between A males and $A + B$ males will in turn be favored because their sons will be superior reproducers, and the female criteria will thus change. As noted by Fisher (1958), this cycle can repeat itself over and over. The likelihood that superoptimal ($A + B$) stimuli do indeed exist at each stage is high, even though previous selection on females favored their using A as a cue. This is because animal nervous systems are interconnected in many complex ways, and it is very unlikely that the parts controlling a female's "decision" process are completely isolated from the influence of all receptors other than those for stimulus A (see Stein and Hart 1983 for evidence of the interconnectedness of control centers in mammalian brains).

An additional cause of change is that selection on females to favor males with novel stimuli (B in the example above) can result in changes in female neural properties such as number and strength of stimulatory and/or inhibitory connections, and sensitivity of receptors so as to increase the female's ability to sense or respond to the new stimulus (B). Thus if females evolve from criterion A to criterion $A + B$, the relative importance of stimulus A can change because, in effect, the female becomes "rewired." The amount of change in the importance of A is unpredictable and could even involve its eventual elimination as a signal.

Animal neurons have two well-known properties that could result in selection favoring continued "embellishment" of male stimuli. Summation, the additive effect of piling on different stimuli in eliciting a response, could result in selection favoring increased complexity of male stimuli. And habituation, the usual tendency of responses to decrease with continued or repeated stimulation, could favor males producing novel or alternative stimuli (a similar argument with respect to precopulatory courtship was made by Jackson 1981; it is discussed at length by West-Eberhard 1984).

It follows that the female criteria are not expected to remain constant but to slip and slide erratically across the phenotypic landscape, following chance occurrences of male variants that increase female receptiveness. The unpredictable nature of this evolution has a very important consequence. At any given point in the evolution of a species, there are undoubtedly many possible ways in which further change could occur; in different, isolated populations it is unlikely that exactly the same combination of changes will occur, especially when stimulus complexity is high. The result is that evolution is likely to be divergent as well as rapid.

Effects of female choice on genitalia

Runaway evolution of genitalia as opposed to other characters is likely to be unusual in one important respect. Runaway sexual selection on characters like bird plumage can continue until the character being selected by females becomes limited by natural selection. For instance, plumage eventually becomes so extravagant that its disadvantage in other contexts, such as predator avoidance, counterbalances its advantage in sexual contexts (Fisher 1958; Dominey 1983). But in the case of genitalia, the "braking" effects of natural selection are probably very weak. In rare instances elaborate genitalia may result in a direct disadvantage for males, as for instance in the small male spider *Tidarren* (Fig. 5.3), which has very large pedipalps; it routinely breaks off one of the pedipalps, presumably to make it easier to move about (Gertsch 1949). Generally, though, the main cost to males of species-specific genitalic modifications is simply that of the generally small amounts of material involved. This implies both that weak female choice is sufficient to induce evolution and, perhaps more important, that runaway evolution of genitalia will be less readily checked than that of many other characters.

Sexual selection by female choice differs in an important way from the other hypotheses already discussed in that the selective context in which it can occur is much more common in nature. In effect it can occur every time a male and female copulate. This contrasts with the species isolation hypotheses, in which selection operates only when all precopulatory isolation mechanisms have failed, and the mechanical conflict of interest hypothesis, in which selection operates only when male and female interests associated with the mechanical aspects of copulation are in conflict. In addition, sexual selection is a comparative process that depends on phenotypic frequencies in the population; a male with a trait that confers relative success at a given moment will be only average in a later population. There are no "final" solutions to adaptive problems. Sexual selection by female choice is thus expected to be common and unrelenting in its action on genitalia.

Fisher visualized the runaway process as beginning with natural

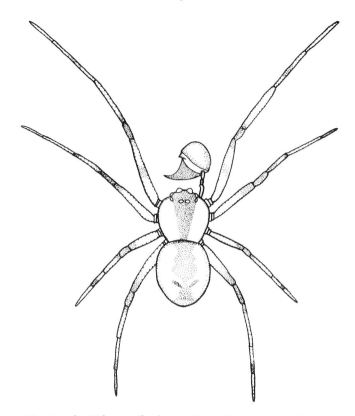

Figure 5.3 A male *Tidarren fordum* spider, showing oversized genitalia. The male pedipalps are so large that the male routinely breaks one off soon after he matures, presumably to make it easier to move around in search of females. Aside from occasional extreme cases like this, genitalia probably represent only a slight cost for males. This means that natural selection is less likely to check the evolution of genitalia than that of other characters such as extravagant plumes and bright colors, which are thought to be subject to sexual selection by female choice. (After Chamberlin and Ivie 1943.)

selection rather than sexual selection on females to discriminate among males on the basis of some character (however, see Lande 1981). That is, by discriminating among males the female originally derives some benefit other than having especially stimulating sons. West-Eberhard (1984) specifies two contexts in which male signals used by females evolve: "successful propaganda," meaning that the female uses a male character to judge mate quality or to cue her reproductive processes, and males evolve to emphasize this character; and "sensory trap," meaning that males use preexisting female receptors and responses that have arisen as a result of selection in other contexts (such as predator avoidance) and emphasize those aspects that the females are preadapted to respond to. Both types of evolution could occur in genitalia. As shown in the next chapter, several female reproductive processes, including egg maturation, sperm transport, tendency to remate, and others, are often triggered by stimuli associated with copulation. This is not surprising, because it allows females to avoid inefficient expenditure of energy and materials. Natural selection must often favor females that accurately sense stimuli associated with copulation and use this information to cue their reproductive processes. These naturally selected traits could then give rise to sexual selection and to Fisher's runaway process.

Sexual selection by female choice could also result from the evolution of special sense organs on male genitalia, as found in insects (Spielman 1966; Peschke 1979), poeciliid fish (Clark and Aronson 1951; Gordon and Rosen 1951), and nematodes (Croll and Wright 1976; Wright 1978). Such organs could improve a male's chances of depositing sperm in an appropriate site or at an appropriate moment when it is more likely to be transported by the female (that is, natural selection could favor the organ). Females that gave preference to males possessing such organs would then be favored because they would have superior sons, and sexual selection could ensue. Thus the stimuli provided by the sense organ itself or its mechanical configuration could become the basis for the original, naturally selected female preference which gave rise to a runaway evolutionary sequence.

Female discrimination could originate in still another context. In a number of species spermatophores may provide nourishment to the female (Friedel and Gillott 1977; Boggs 1981; Gwynne 1983; see also Thornhill and Alcock 1983). Natural selection on females to insure adequate nutritive value could favor discrimination in favor of larger or more numerous spermatophores. Boggs (1981) has shown that this may occur in the butterfly *Dryas julia*, as the female's tendency to remate is negatively correlated with the size of the spermatophore received in her first mating. Males of the grasshopper *Melanoplus sanguinipes* transfer an average of seven spermatophores per copulation, even though a single spermatophore contains enough sperm to fertilize several clutches of eggs (Friedel and Gillott 1977). Runaway sexual selection would probably be "restrained" in this type of situation, however, since selection should favor females that accurately discriminate against deceptive cues and in favor of spermatophores with greater nutrient value.

Finally, runaway selection might be set off by female responsiveness arising by drift without any intervening naturally selected advantage (Lande 1981). I would argue, however, that the strong tendency for male intromittent organs rather than other structures to undergo runaway evolution is evidence against the importance of drift; as will be shown in detail in Chapter 7, female response to male genitalia must often be favored by natural selection.

Mechanical fit of genitalia

Females could use two different types of criteria to discriminate among male genitalia. The first, illustrated in the discussion above, is the stimuli she receives from his genitalia. A second type of criterion depends on the mechanical fit between male and female genitalia. For instance, females could discriminate among males on the basis of their ability to hook into a particular cavity, to turn a given corner, to brace themselves in a way allowing them to twist past a given prominence. This aspect of the sexual selection hypothesis is similar to that of the lock and key hypothesis. The important difference is that selection for species isolation, which the evidence in Chapter 4 shows is unlikely, is replaced by selection due to female choice.

A concrete example of possible mechanical discrimination occurs in the springhaas (*Pedetes surdaster*). Both the penis and vagina are unusually long, and when the dorsal cervical connective in a dead female is pressed at the point where the male penis bone (baculum) lies during copulation, the cervical ostia become slightly dilated, which would presumably allow sperm to enter (Coe 1969); the penis of this species probably also functions in other ways, as indicated by the spines on the glans and an inflatable terminal membrane. Other species in which mechanical "gate opening" may occur include the horse, in which the glans becomes greatly inflated at the moment ejaculation begins (Walton 1960), presumably to open the cervical canal (Hunter 1975), and the cixiid homopteran *Vincentia*, in which parts of the aedeagus dilate the female genital chamber to allow insertion of the spermatophore (Fennah 1945).

This reasoning can also be applied to the coevolution of male grasping organs and female structures in some insect groups, including odonates (Corbet 1962), cicindellid beetles (Freitag 1974), and colletid bees (Toro and de la Hoz 1976). Females that mechanically favored those males better able to grab them (for instance, by having depressions to receive an extra tooth on the clasper organ that reduced slippage) would have an advantage because such males would be less likely to be dislodged by other males or by the female's own movements. It would also explain why females of such species have *depressions* of one sort or another that aid the male in holding on, rather than the anticlasper devices predicted by the mechanical conflict of interest hypothesis (see Fig. 4.3). The result would be a hand-in-glove fit between male and female structures, even in the absence of selection for species isolation.

Females could also discriminate on the basis of a combination of mechanical and stimulatory properties. For instance, the different parts of the male genitalia of araneomorph spiders serve a variety of mechanical functions for bracing the male genitalia and driving the intromittent structure (embolus) into the often complex ducts leading to the spermatheca (Gering 1953; van Helsdingen 1965; Grasshoff 1973; Blest and Pomeroy 1978; see also Fig. 2.1). The genitalia generally lack stimulatory structures such as bristles or brushes of hairs, but they could turn out to be complex mechanisms with the ultimate function of delivering a single stimulus — the embolus entering the bursa or spermatheca and perhaps stretching it, as occurs in the spider *Agelenopsis* (Gering 1953).

The two modes of discrimination, stimulatory and mechanical, would lead to different patterns in the evolution of male versus female genitalia. When females discriminate on the basis of stimuli, the differences in their criteria would often be manifest mostly in their nervous systems and thus would be invisible in their external morphology. This would result in diverse male genitalic morphologies but uniform female morphologies, a common pattern which I pointed out in Chapter 3. On the other hand, mechanical discrimination would involve changes in female structures and would result in concomitant evolution of both male and female structures, as seen in many spiders (Kaston 1948), *Cicindela* beetles (Freitag 1974; personal communication), and other groups.

A combination of both mechanical and stimulatory criteria have recently been demonstrated in damselflies (Fig. 5.4). The males' abdominal appendages, often species-specific, are used to grasp females (see Table 11.1). In some species it has been shown that

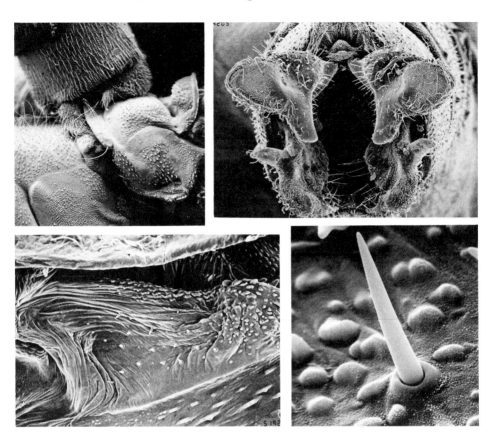

Figure 5.4 The species-specific male grasping organs (abdominal appendages) and the species-specific area on the female prothorax that they hold, in the damselfly *Enallagma glaucum*, illustrating the sense organs that enable the female to discriminate among males on the basis of their grip; a female gripped by a male with claspers that are not typical of her species refuses to mate (she does not bend her abdomen forward to make genitalic contact). *Upper left*, lateral view of the male's abdomen (above), holding the dorsal surface of the female's thorax. *Upper right*, posterior view of the male appendages splayed out laterally as they are while clasping the female. *Lower left*, dorsal view of the area on the female where a male appendage clasps, showing the sensilla (short, thick hairs). *Lower right*, single sensillum magnified. This trailbreaking study showed that the female receptors have species-specific forms that correlate with the species-specific forms of the male appendages. (From Robertson and Paterson 1982.)

males are mechanically unable to grasp females of other, sympatric species (Paulson 1974). The male appendages seem ill-equipped to stimulate females, as they are apparently immobile once they have seized the female (personal observation) and generally lack spines or brushes. Instead they fit in a hand-in-glove way with indentations on the female head or thorax (Corbet 1962; Johnson 1972; Paulson 1974). Nevertheless, the females of three genera in two different families are known to differentiate among males on the basis of stimuli received from these appendages; a female will not bend her abdomen forward to bring her genitalia into contact with the male's secondary genitalia and copulate when she is seized by a male of another species or by a male of her own species whose clasping organs have been modified (Loibl 1958; Krieger and Krieger-Loibl 1958; Robertson and Paterson 1982; Tennessen, in Tennessen 1982). Females of the damselfly genus *Enallagma* have mechanoreceptors distributed in species-specific patterns on an area of the thorax where species-specific male organs contact them (Robertson and Paterson 1982).

Finally, sexual selection on the visual appearance of male genitalia, was documented by Wickler (1966, 1969) for primates, and may also occur in some lizards (Bohme 1983). Some of Wickler's claims seem not to be true however, at least for humans (Short 1979). If so, this leaves selection for tactile stimulation the most likely selective factor explaining human males' distinctive and oversized genitalia.

Genitalia and overall male fitness

Females could possibly sense other aspects of male quality, such as size, strength, and vigor, through genitalic size or stimulation, but I think this is not likely to occur often. Correlation between male body size and genitalic size was investigated in ten species of arthropods — seven insects, two spiders, and one milliped (A. Alvarado, J. Arias, C. Arias, D. Briceño, W. Eberhard, R. Lopez, O. Rocha, and E. Rojas unpublished data). In only five species was a measurable part of the male genitalia significantly correlated with at least one measure of body size (such as pronotum width or wing length; see Fig. 5.5). So male genitalia may not be consistently good indicators of male size, at least in arthropods. Presumably this correlation could give some species another naturally selected advantage for those females able to discriminate among males on the basis

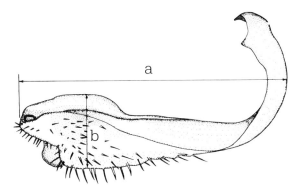

Figure 5.5 Gonopod of a male *Nyssodesmos python* milliped. Dimension *a* correlates significantly with both head width and dorsal tergite width, and females could presumably sense a male's size by the size of his genitalia. Such correlations are weak or nonexistent in some arthropods.

of their genitalia, but it would not so easily result in a runaway process, as there would be natural selection on females to detect "braggarts" with relatively oversized genitalia. There is no obvious reason to believe that genitalic characters generally correlate particularly well with other aspects of overall male fitness, and many additional characters that are better indicators of overall fitness are often available to females.

Major evolutionary patterns that are explained

A logical consequence of the female choice hypothesis is that intromittent genitalia, rather than other organs, should evolve rapidly and divergently because intromission necessarily entails stimulation of the female in her genital region and a certain degree of mechanical fit. In a sense intromission preadapts females to use the properties of male genitalia rather than those of other organs to cue a series of physiological processes that determine the fate of a male's sperm once it is inside her. By the same token, in animals with external fertilization and in those with internal fertilization that use secondary rather than primary genitalia in copulation, the primary genitalia would be expected not to follow the general pattern of rapid evolutionary divergence in genitalic structure; the data in Tables 1.1–1.4 clearly fit these predictions.

It also follows from the female choice hypothesis that in many groups female genitalia will vary less than male genitalia; the common occurrence of this pattern was cited in Chapter 3 as evidence against the species isolation hypotheses. Female discrimination among males on the basis of genitalic stimuli is probably often manifested in the females' nervous systems rather than in their morphology and is thus invisible externally. On the other hand, when the female criterion is mechanical rather than or in addition to stimulatory, one would expect concomitant evolution in male and female genitalia, so a mixture of patterns in the evolution of female genitalia is predicted; again, this is what occurs (see Chapter 3).

Another general pattern, difficult to document precisely but probably real, is that in groups with structurally more complex genitalia, species are more often distinguishable by genitalic characters than in groups with simpler genitalia (Dobzhansky 1941; on crickets, R. D. Alexander, manuscript). Complexity is difficult to quantify, and it is difficult to take into account the fact that different groups have been separated for different lengths of time. In addition, genitalia and other characters have been studied with varying degrees of care in different animal groups, and it is probably easier for the human eye to detect subtle differences in the form of a genitalic structure that is in the midst of other complex genitalic components that provide points of reference (H. W. Levi, W. Shear, personal communication). The pattern is thus difficult to demonstrate conclusively, but I think it is real. For example, in several large groups of arthropods — most amphipods, termites, aphids and coccid homopterans, some crickets, and many parasitic wasps — the male genitalia are quite simple, and they generally fail to show interspecific differences (Bousefield 1958; Alexander and Otte 1967; Askew 1968; Tuxen 1970). In some other groups, like antrodiaetid and *Tetragnatha* spiders, relatively simple genitalia do show interspecific differences (Coyle 1971; Levi 1981), and more careful studies in the future will undoubtedly show this to be true in other groups. But the converse — complex genitalia that do not vary among the species in a sizable group — is notably rare.

Sexual selection by female choice explains this trend. Different species are thought to diverge because of the improbability that female choice will focus on exactly the same sets of male characteristics in different populations; the larger the number of characters to choose from, the greater the likelihood of divergence.

The "extravagance," or apparently superfluous complexity, of many genitalia was noted in Chapter 1; as I suggested there, this pattern is also an expected result of sexual selection. The arbitrary nature of females' criteria and the potentially cumulative nature of complex modifications mean there is no predictable upper limit to genitalic complexity.

A final probable trend should be mentioned: the frequent association between elaborate premating courtship and relatively simple and uniform genitalia I mentioned in Chapter 2 (see Table 2.1, Fig. 2.2). I argued there that the existence of so many clear exceptions cast doubt on the previous interpretations of the data as supporting the species isolation hypotheses. It may be, however (though with many exceptions) that groups in which males and females exchange complex premating signals tend to have less complex and more uniform genitalia. An imperfect correlation of this sort is predicted by the sexual selection hypothesis. Females of species that discriminate among males on the basis of premating signals have already discriminated strongly using nongenitalic cues before copulation begins; they may sometimes use additional genitalic criteria and sometimes not. (See Thornhill and Alcock 1983; West-Eberhard 1983, 1984, for reviews of evidence that premating signals are often sexually selected.) The arbitrary nature of both the type and number of cues that females will use makes correlation or lack of it impossible to predict for particular cases, but one would expect a loose correlation like that observed. In sum, a strong correlation between complex premating signals and simple male genitalia is predicted by the species isolation hypotheses and does not occur; a weaker correlation is predicted by the sexual selection hypothesis and probably does occur.

Relationships to other hypotheses

It is important to clarify the relationships between the female choice hypothesis and both the species isolation hypotheses and the mechanical conflict of interest hypothesis. Fisher (1958) noted that selection favoring female choice could sometimes arise because of the advantage of avoiding interspecific matings and could then subsequently "run away." Thus selection for species isolation could sometimes lead to sexual selection by female choice. But runaway evolution by female choice can also be initiated in many other contexts, as discussed above, so female choice does not require prior selection for species isolation.

Conflict of interest between male and female can be construed to cover a range of phenomena, including sexual selection by female choice. The conflict in this case stems from each male attempting to fertilize all of a female's eggs, while the female attempts to allow only certain males access to her eggs. There is, however, an important difference between the hypotheses. Sexual selection by female choice favors females that reject some males in order to favor others. The conflict of interest ideas of R. D. Alexander (manuscript), Lloyd (1979), and Wing (1982) suppose that the female gains from mechanically rejecting any and all males (at least sometimes) in order to promote her own survival and/or reproduction — ovipositing, feeding, and so on. Thus one hypothesis entails selective rejection; the other, wholesale rejection. I argued in Chapter 4 that female mechanisms for indiscriminate rejection are likely to be evolutionarily conservative and are thus unlikely to explain genitalic evolution, and that such rejection, which undoubtedly does occur sometimes in nature, usually occurs behaviorally prior to genitalic contact.

I should clarify that both Alexander and Lloyd wrote in general terms, and I do not suppose that either would deny that conflict of interest concerning genitalic morphology could occur in the context of sexual selection by female choice. However, both authors cited only examples of what I have termed mechanical conflict of interest, with the male physically forcing his way past female defenses. Mechanical conflict of interest, as I understand it, has some serious logical difficulties in explaining genitalic evolution (see Chapter 4) and generates different predictions, including concomitant changes in female and male genitalia, and anticlasper structures in females. For these reasons it seems important to distinguish it from sexual selection by female choice.

6 *Male-male genitalic competition*

Sexual selection encompasses both selection by choosy females and male-male struggles for access to females. In Chapter 1 I noted that genitalia might offer an unusually "pure" case of sexual selection by female choice because males very seldom use their genitalia in aggressive interactions with other males. The clearest exception I know occurs in some male primates, which threaten other males by using their brightly colored erect penes as aggressive signals (Wickler 1969); in this exceptional case male genitalia are diverse and may well have evolved their species-specific forms and colors as a result of sexual selection. (See Darwin 1871 and West-Eberhard 1983, 1984 for many other, similar examples of rapid divergence in signaling structures.) Male-male genitalic competition could also occur in another context, however — inside the female. This chapter examines the possibility that male genitalia tend to diverge rapidly because of sexual selection resulting from male-male interactions both on the surface of and within the female. The overall conclusion is that such competition may occur in some cases, but that it is highly unlikely in many others in which genitalia are nevertheless species-specific in form.

First, it is necessary to clarify the relationship between female choice and male-male competition with respect to sexual selection on genitalia. Both processes have the same result: some males have greater reproductive success than others. It might seem easy to distinguish the two processes, since male-male conflict does not involve females directly, but females can, in effect, choose the winners of male-male contests by accepting as mates only the winners of such interactions. They may do this by mating only after a long

Figure 6.1 A tangle of male *Centris pallida* males attempting to copulate with a newly emerged female. Even momentary failure by the female to respond appropriately to a given male could result in that male's displacement by a competitor. (From Thornhill and Alcock 1983).

courtship that is likely to attract the attention of nearby males, by mating only with dominant males where males have aggregated, and so on (see Thornhill and Alcock 1983). The criteria in such cases of "passive" female choice (Thornhill and Alcock 1983) would be the same as those that determine success in male-male combat. A possible example is the bee *Centris pallida* (Fig. 6.1). If an emerging female copulates with a male that is unable to maintain his position on her back long enough to perform postcopulatory stimulation (usually involving one or more "superfluous" genitalic thrusts), she will mate again with the male that replaces him; male genitalia in *Centris* are species-specific in form (J. Alcock, personal communication).

It is not always possible to tell the difference between male-male competition and female choice. In the case of male-male competition within the female, the discrimination is especially difficult because the female's own body is seen as the "battlefield" on which the competition is carried out, and changes in female morphology, physiology, and behavior obviously could affect the outcomes of such conflicts. Thus the possibility of female choice lurks behind the types of male-male conflict discussed below. The difficulty in resolving this point is a recurrent theme in this chapter.

Sperm displacement

One way in which males could use the mechanical properties of their genitalia to compete with each other is by interfering with the sperm that other males have already deposited in females. For example, a male might use his genitalia to remove other males' sperm from the female, or pack them away from sites that give access to eggs (Waage in press). The removal function has actually been demonstrated (see Fig. 2.3) in some damselflies (Waage 1979). (Sperm plugs in other species that may prevent subsequent males from obtaining fertilizations are discussed in Chapter 10; they are not known to be elaborate and species-specific in form and are thus not pertinent here.)

Is it possible that the tendency for genitalia to evolve rapidly and divergently usually results from direct male-male competitive interaction to manipulate other males' sperm within females rather than from female choice? The answer, at least in general terms, is a fairly definite no. In many animal groups in which male genitalia are elaborate and species-specific, male genitalia do not reach the areas in the female where sperm from previous matings is stored (see the section on female topology later in this chapter). A particularly clear example is the Lepidoptera in which species-specific male genitalia are very common; the bursa copulatrix of the female (*b* in Fig. 6.2), where the male's genitalia rest during copulation, is connected to the spermatheca by a long duct (Fig. 6.2). The intromittent organ is physically separated from the sperm storage site in some other insects (Davey 1965; see also Fig. 11.1), poeciliid fish (Rosen and Gordon 1953), snakes (Saint-Girons 1975), many mammals (Walton 1960), many molluscs (Duncan 1975; Beeman 1977; Wells and Wells 1977; see Fig. 7.1), chaetognaths (Ghirardelli 1968), and other groups. It is just not physically possible for males of these groups to use their genitalia to mechanically remove, displace, or supersede previous males' ejaculates. In addition, the spermatophores of a number of groups are structurally complex and have apparently evolved rapidly and divergently (see Table 1.3). The details of their functioning are not well known, but what is known about those that have been studied, such as the explosive spermatophores of pseudoscorpions (Weygoldt 1969), scorpions (Francke 1979), and cephalopods (Wells and Wells 1977), suggests that they drive their own sperm *into* the female rather than extracting the sperm of others from her.

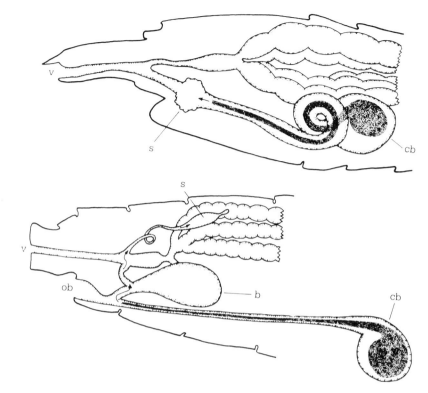

Figure 6.2 Above, ancestral and, *below,* derived forms of female genitalic topography in Lepidoptera (stippled bodies are spermatophores). The ancestral form has a single opening (*v* = vulva) through which sperm are introduced and eggs are laid; in the derived form copulation occurs in a separate opening, the ostium bursae (*ob*). Sperm released from the spermatophore in the derived form must pass through a series of connecting tubes (arrows) before arriving at the spermatheca (*s*). Male genitalia, which are commonly species-specific in form, reach only the bursa (*b*) and corpus bursae (*cb*) during copulation and thus cannot physically interfere with sperm stored in the spermatheca from previous matings, so direct male-male genitalic competition within the female is not possible. (After Common 1970.)

In some instances male genitalia *are* used to displace other males' sperm, as Waage (1979, in press) has demonstrated in some damselflies and possibly also some dragonflies. Another probable case, judging by the morphological studies of Leigh-Sharpe (1920, 1922) occurs in sharks and their relatives. Here the male intromittent (clasper) organ is actually a basal element of the pelvic fin that is rolled up like a scroll to form a tube. Two other tubes empty into the proximal end of the clasper tube: the cloaca, through which the sperm travel, and the "siphon," which in the dogfish is a hollow, blind-ended muscular tube filled with seawater. When the siphon contracts, water shoots out of the distal tip of the clasper, and a structure at the tip gives this spray a "spiral rotatory motion." Leigh-Sharpe supposed that the purpose of this was to spray sperm into the female; however, given the seawater content of the siphon, the extraordinary adaptation to spray all sides of the female's duct, and the "enormous dimensions" of the muscular siphon, which occupies a substantial fraction of the male's body cavity in some species of *Galeus* and *Mustelus,* it seems more likely that the male uses the siphon to wash out the female's lower reproductive tract with seawater (to give her a contraceptive douche) before depositing his own sperm. There is probably even more to this particular story; the muscular siphon of some skates *(Raia),* instead of being hollow, is nearly filled with a gland of undetermined function. Does it produce spermicides?

There are of course, a number of other ways in which males' ejaculates might compete directly with each other inside the female. For example, the semen might include chemicals that kill or inactivate other males' sperm or that induce the female to discard them; some males might produce larger ejaculates that swamp those of other males; and in species with dimorphic sperm, some types of sperm might be adapted to reduce the chances of other males' sperm being used for fertilization (Silberglied et al. 1984). All of these types of competition could occur without affecting the morphology of the male (or the female) genitalia and would thus pass unperceived in morphological studies. At least some of these mechanisms probably do exist — the evidence I have seen is suggestive but incomplete — but they are peripheral to the morphological questions being examined in this book.

Although it is unlikely that male-male genitalic competition to displace ejaculates has accounted for rapid genitalic divergence in many groups, the question remains, how important has it been in those groups in which such competition occurs? Our current state of ignorance about the functioning of different parts of these animals' genitalia makes confident answers impossible. For example, the various tails and flanges on the "scoops" of damselfly penes (Kennedy 1919, 1920; see Fig. 2.3) could be especially effective devices for cleaning out hard-to reach sperm, or they could be stimulatory structures mounted on the basic scooping structure. In fact, they could be both, and the stimuli normally associated with sperm removal could have been the criterion first used by females to discriminate in favor of better sperm-removing males, subsequently giving rise to a bout of runaway evolution. Thus direct male-male competition involving sperm displacement could have been responsible for rapid divergence in odonates' genitalic morphology, but it is also possible that female choice was an important selective factor.

Clasping devices

Another way in which males might use their genitalia in male-male competition is via clasping devices that hold the female and thus prevent other males from obtaining access to her (Fig. 6.3; see also Figs. 4.2, 4.3). Thornhill and Alcock (1983) discuss this possibility in insects, and they note that in addition to clasper organs, a variety of other male genitalic characteristics (length, presence of spines, inflatable components, or other interlocking features) could also function in this way. If some males of a species hold females more effectively than others, and if some at least occasionally dislodge less effective males before they complete copulation, then sexual selection by male-male conflict could act on the males' genitalic holdfast devices and result in rapid evolution.

Is it possible that this type of competition has been responsible for the general trend toward rapid divergence in genitalic evolution? As in the previous section, I think the answer is an almost certain *no* in general terms, but a clear *maybe* in some specific cases. In a high percentage of cases the necessary conditions for the hypothesis do not occur. In some groups the species-specific male genitalic structures do not even remotely resemble holdfast devices or are not used to physically hold the male close to the female: examples include species of bumblebees (Richards 1927a), lepidopterans (Lorkovic 1952; Scott 1978; Platt 1978), and beetles (Alexander 1959; Trejos and Eberhard, unpublished; see Chapter 10). In many other groups with species-specific genitalia, the male uses other appendages to hold the female. This occurs in rodents; many insects, including odonates (Corbet 1962), beetles (Wocjik 1969), wasps (Eberhard 1974), and bees (Alcock and Buchmann, in press); some cephalopods (Wells and Wells 1977); pseudoscorpions (Weygoldt

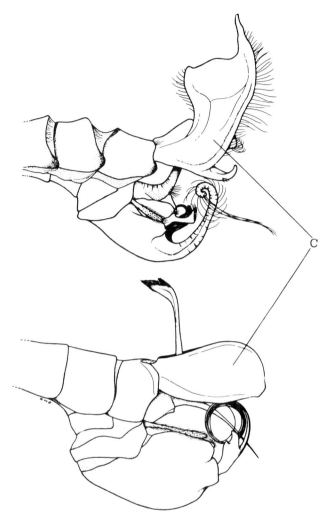

Figure 6.3 Male terminal abdominal segments of two species of *Bittacus* scorpionflies, showing dramatic differences in the massive clasping organs or epiandrial lobes (*c*), as well as in a variety of other structures. If rapid divergence in genitalic evolution were the result of male-male competition to clasp females more securely, one would predict that divergence would occur in the clasping structures but not in the other genitalic structures as seen here, since such powerful claspers would presumably make the holding properties of other structures comparatively insignificant. (From Byers 1970.)

1969); scorpions (Kaestner 1968); millipeds (Haacker and Fachs 1970; Haacker 1979; personal observation); and spiders (Robinson and Robinson 1980). It is unlikely that the mechanical properties of these males' intromittent genitalia have an appreciable effect on the strength of their hold on females.

In still other species it is highly unusual for more than one male to encounter a female at one time. In many species males with species-specific genitalia sequester females from other males before or during copulation: examples include dung beetles (Howden 1979; Haffler and Matthews 1966), and crayfish and lobsters (Nelson and Hedgecock 1977). Probably in many other species the typical population densities are so low, and/or mating is so rapid, that the likelihood of a second male coming upon a pair *in copulo* is quite low.

Finally, in some species with species-specific genitalia males achieve such secure genitalic locks with females during copulation that other males seem unable to break them apart (Thornhill and Alcock 1983), and thus the selective advantage of further evolution of increased locking ability (and thus divergence of structures) seems to be lacking. Essentially unbreakable locks seem to occur, for instance, in the scarab beetle *Ceraspis brunneipenis* (personal observation; male genitalia are species-specific in this genus, Frey 1973): males have an array of large saclike structures that are inflated during copulation and make it very difficult to separate copulating individuals. Pulling them apart by hand requires much more force than an individual beetle is likely to be able to produce, and sexually active solitary males, which will mate readily with solitary females, behave as if pairs are not separable, making no attempt to dislodge other males they encounter *in copulo*.

For these reasons, sexual selection stemming from direct male-male competition involving holdfast devices is not of general enough significance to explain the widespread tendency for genitalia to evolve rapidly and divergently. This idea is not ruled out for specific cases, but again, it cannot be distinguished in these cases from the possibility of sexual selection by female choice. Because the female herself is physically interposed between competing males, it is often both possible and selectively advantageous for her to prejudice the competition.

Consider, for example, a population in which some but not all males possess a new, superior holdfast device that increases the probability of not being dislodged by other males by, say, 30 percent. Females who are able to discriminate in favor of the superior males (that is, those females that improve the superior males' chances of fertilization by more than 30 percent) will be favored because they will produce superior sons. Females could discriminate in a number of ways: by obliging the male to resist the attempt of other males to dislodge him or by engaging in particularly energetic dislodgement behavior herself; she could also use stimuli and/or mechanical fit associated with the holdfast device as criteria for allowing the sperm to fertilize her eggs, which could result in a bout of Fisherian runaway evolution. Thus either one or both selective factors (female choice, male-male competition) could be operating in these cases, and it is difficult to discriminate between them.

This example leads directly to a generalization that in a way summarizes the chapter up to this point: only under certain circumstances is direct male-male competition via the mechanical properties of their genitalia possible, and in all of these situations female choice can also occur. On the other hand, female choice among males on the basis of their genitalia is possible in a much greater variety of situations, as will be shown in the next three chapters, in many of which male-male competition is not possible. Thus female choice is more likely to be general enough to explain the range of situations in which rapid divergent genitalic evolution is known.

Male adaptations that short-cut female choice

Still another way in which males could compete within females is by placing their sperm in a position that reduces the possibility of supplantation by the sperm of other males. Thus sexual selection by male-male competition might favor male genitalic structures that circumvent the female characteristics responsible for sometimes preventing their sperm from achieving fertilizations. Although males may seldom be able to accomplish fertilization as a direct and inevitable result of insemination (see Tables 1.1–1.3), there is suggestive evidence (Lloyd 1979 and below) that this kind of competition has indeed occurred in a few groups.

One possible example occurs in homopteran insects in the cicada family. Careful comparative studies of the female reproductive anatomy of several species (Boulard 1965) showed that the structures corresponding to the spermatheca of other homopterans have been transformed into glands that no longer receive sperm. Instead, the sperm are stored higher up in the female reproductive tract in a swelling (ampulla seminalis) in the oviduct wall. In the ancestors of these groups some males presumably achieved precedence for their sperm over that of other males by placing it nearer the ovaries and thus fertilizing eggs within the oviduct before they ever reached the old spermatheca, and later females evolved special pouches to store the sperm of these reproductively superior males.

A similar topographic contest within the female was documented by Drew (1911) in the squid *Loligo paelii*. In this case the variation is intraspecific. The "usual" method of copulation in *L. paelii* is for the male to hold his grenadelike spermatophores near the female's mouth; when the spermatophores explode, the sperm reservoirs attach to the female, and the sperm are carried (probably by female responses, Drew 1911) into receptacles where they are stored, apparently for a long time (weeks or months). When she lays her eggs, the female passes each one past the sperm receptacle, where it is fertilized. If, however, a female is nearly ready to oviposit, males copulate in a different way. The spermatophores are inserted deeper into the female's mantle cavity so that when they explode, the sperm reservoirs are attached near the distal end of her oviduct. There is no special receptacle there, and Drew described the sperm escaping gradually into the water in the mantle cavity "like smoke from a chimney." They are gradually lost from the female to the outside through the funnel. Drew estimated that sperm are discharged for up to about two days within the female after such a copulation. Sperm placed deeper in the female's body (farther up her reproductive tract) probably have first chance at fertilization. They cannot survive for long in such areas, however, so males place them there only when the female is about to oviposit.

Drew pointed out that in some other cephalopods, such as *Octopus*, all spermatophores are deposited in the females' oviducts. Peristaltic movements of the oviduct apparently then transport the sperm in *Octopus*, and they are stored in special "oviducal glands" in the upper half of the oviduct (Wells and Wells 1977). Since in the sister group Nautiloidea spermatophores are attached near the mouth (Haven 1977), deposition higher up in the female tract in squids and *Octopus* is probably derived rather than primitive; again the selective factor probably responsible was male circumvention of internal barriers to fertilization within females. Still other similar cases may occur in the gastrotrich *Dactylopoda baltica*, the cimicid bugs *Haematosiphon* (Fig. 6.4) and *Psitticimex*, the spider mite *Te-*

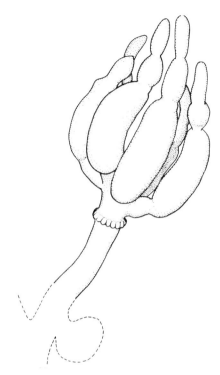

Figure 6.4 An ovary of the cimicid bug *Haematosiphon inodorus*, showing the collar of small enlargements just below the oocytes where sperm are stored. Dotted lines indicate the presumed ancestral condition with a large sperm storage organ farther down the oviduct, which has been lost in *Haematosiphon.*

The site of sperm storage has also been moved "upstream" in the female in other groups; these switches may result from competition among males to give their sperm priority in fertilization. (After Carayon 1966.)

tranychus urticae, where sperm apparently migrate up the female tract to the ovaries (Hummon 1974; Carayon 1966; Helle 1967), and some reptiles in which sperm are stored both in a seminal receptacle and farther up the female tract in the tuba (Saint-Girons 1975). In some insects (for example, the weevil *Anthonomus grandii,* Villavaso 1975), sperm capable of fertilizing eggs are present outside the spermatheca in other parts of the female reproductive tract for up to two weeks after copulation.

Hypodermic or intraperitoneal insemination is a different method of achieving more direct access to fertilizable eggs. The

male inserts his genitalia through the female integument and introduces sperm into her body cavity, where they subsequently migrate to the ovaries and fertilize the eggs. This type of insemination
occurs in a variety of groups, including hemipteran and strepsipteran insects (Imms 1957), perhaps the moth *Malacosoma* (Bieman
and Witter 1982), leeches (Mann 1962), polychaete worms (Schaller
1971; Schroeder and Hermans 1975), at least two groups of molluscs
(Purchon 1977), rotifers (Thane 1974), onychophorans (Schaller
1971), cestodes (Hyman 1951b), gnathostomulid worms (Sterrer
1974), and some turbellarian flatworms (Henly 1974). In some
groups in which hypodermic insemination does not occur, the female body cavity seems to be a surprisingly benign environment for
sperm. For instance, Rowlands (1958) noted that in studies of
chickens, pigeons, cattle, and guinea pigs, intraperitoneal insemination gave just as high fertilization rates as intravaginal insemination.
In some mammals transperitoneal migration of sperm may occur
frequently, as has been documented in humans and sheep (Blandau
1969). Stephenson (1930) noted that sperm also penetrate into the
body cavity in some oligochaete worms. In cimicid bugs, sperm not
only move within the spaces between cells but also move *through*
cells (Carayon 1966)!

Lloyd (1979) discussed hypodermic insemination in insects and
noted that it probably represents a mechanism for sperm precedence. The best-studied case is that of bedbugs (Cimicidae) and their
relatives (reviewed by Carayon 1966 and 1975, from which the
following accounts are taken). In some genera the male deposits
sperm in the female spermatheca, but they later escape into the
body cavity and migrate to the ovaries, where they fertilize eggs. In
other genera the male uses his hypodermic genitalia to inject sperm
at a variety of different sites on the female's body, and the sperm
migrate via the hemocoel and the lining of the oviduct to the ovaries.
In some genera, which are thought to be more derived, the male
inserts his genitalia only into certain specialized regions (ectospermaleges) of the female body (Fig. 6.5), and the sperm that are injected are received in a special internal structure (mesospermalege)
and then migrate through the walls of this structure to the oviduct
and then to the ovaries. In the most derived forms of the mesospermalege, there are structures that direct or even conduct the sperm
to the area where they penetrate the oviduct. In effect, these females have invented a new set of genitalia!

In an extraordinary parallel, some leeches have abandoned the

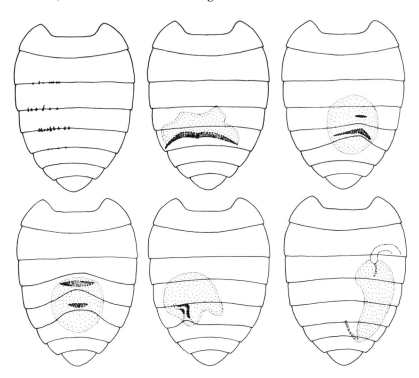

Figure 6.5 Positions and forms of "spermalege" organs of female Cimicidae (bedbugs and their relatives) in different genera. Males use hypodermic insemination, and females have evolved a new set of genitalic structures (paragenitalia) that receive the males' genitalia and semen. The "ectospermalege" (dense stippling with dashed borders) receives the stabbing insertions of the male genitalia, while the internal "mesospermalege" (light stippling) receives the injected semen and destroys part or most of the sperm before they escape into the hemocoel and migrate to the oviducts and then to the ovaries. *Top left,* the genus *Primicimex* has no spermalege, and scars from copulations (black) show that males insert their genitalia in a number of different sites. Large numbers of sperm reach the ovaries in this genus. *Lower right,* in the genus *Stricticimex* both the ectospermalege and the mesospermalege form conducting devices that route the sperm directly to the oviduct, killing a large portion of them in the process. (After Carayon 1966.)

indiscriminate hypodermic insemination that is characteristic of most other leech families; in the genus *Pisicola* the female reproductive system (all leeches are hermaphrodites) has a special "copulatory organ" where the hypodermic spermatophore is placed (Fig. 6.6), and fibrous connective tissue links this site directly to the ovarian sacs (Mann 1962). Strepsipteran insects are still another group with hypodermic insemination in which females have secondary genital pores into which males introduce their genitalia (Imms 1957).

These examples highlight a question regarding genitalic evolution that also arises in other examples in this section but that has not been touched upon. It may be true that males evolved hypodermic insemination to increase their chances of obtaining fertilizations, but why did females then evolve a new set of receiving and conducting structures for sperm? If the line of reasoning being pursued here is correct, females may have been selected to favor fertilizations by males that inject sperm into a particular area of their body,

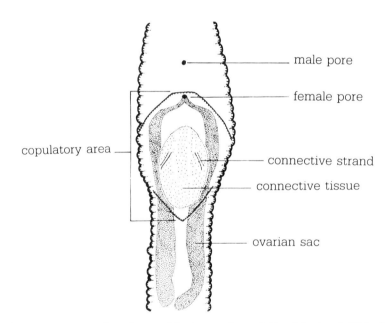

Figure 6.6 External, and part of the internal, genitalia of the leech *Pisicola*, showing an extraordinary convergence to the bedbug paragenitalia illustrated in Fig. 6.5. The leech is a hermaphrodite, with both male and female openings; only the female internal organs, which lie in the body cavity posterior to the female pore, are shown. After the partner's hypodermic spermatophore is inserted into the copulatory area, the sperm migrate to the "connective tissue" mass below, then move via the connective strands to the ovarian sacs. It is not known whether or not sperm and/or seminal fluid are degraded in any of these tissues, as occurs in the bedbugs. (After Mann 1962.)

or they may have been selected to confine sperm in order to effect subsequent control of paternity (on the basis of information regarding male quality gained during courtship, for instance) rather than allow immediate fertilization.

This second possibility is supported by several details of the apparent evolution of female paragenitalia in cimicids. Carayon (1966, p. 129) notes that female adaptations "largely resemble defense reactions," such as local integumentary hypertrophies (presumably to minimize the trauma resulting from hypodermic insemination), concentrations of amoebocytes, secretions of PAS-positive membranes, and so on. In all species with hypodermic insemination there is also a "syncytial body" that blocks each ovariole posteriorly (the direction from which sperm arrive), and Carayon has shown that this body resorbs sperm flowing into the ovariole. Primitive mesospermaleges (internal sperm-receiving organs) filter out seminal components that activate the sperm, and Carayon believes that the mesospermalege thus prevents or retards sperm activation. In more derived forms some of the sperm are actually destroyed in the mesospermalege by amoebocytes. In the most derived mesospermaleges *(Crassicimex, Stricticimex)* the sperm that are not destroyed are stored and conducted slowly, via a "conductor cord," where still another portion of the sperm is resorbed before they reach the oviduct; "only a few spermatozoa leave the conductor cord intact" (Carayon 1966, p. 129), and even these must still pass through the sperm-destroying syncytial body before arriving at the ovarioles. This derived condition contrasts with the intense and prolonged spermathemie (sperm in the blood), distended storage organs packed with sperm, and the "massive migrations toward the ovarioles" (p. 128) that occur in less derived forms such as *Primicimex*, which lack a mesospermalege.

It is possible that resorption of sperm and seminal fluid results in nutritional benefits to females, but Mellanby (in Carayon 1966) carefully tested the effects of repeated matings on female fecundity in *Cimex* and showed that increased matings had no effect other than an increase in female mortality. In addition, even if the semen do provide appreciable nutritional benefits in some species, the adaptive advantage of having multiple organs for destroying sperm and seminal fluid is unclear, since presumably even in forms like *Primicimex*, which lack most of these structures, the semen is eventually degraded and resorbed by the female near the ovary.

Another line of evidence suggesting the possible importance of female defense against males involves the external sites on the female body (ectospermaleges) that are specialized to receive male genitalia. These seem to function, in the most primitive state, to minimize female injuries resulting from males' genitalic insertions (they are sometimes fatal). In more derived forms it is clear that these ectospermaleges conduct the sperm to the internal mesospermalege, and conducting structures have originated independently at least twice. The sites of the ectospermaleges vary dramatically among genera, and the sites of insertion are not precisely concentrated on a given spot within the ectospermalege (see Fig. 6.5). This rapid shifting of intromission site, also seen in the mite family Rhodacaridae (Fig. 6.7; Lee 1970), could be the result of an evolutionary race between males and females with conflicting interests, as envisaged by Lloyd (1979), with males shifting injection sites to avoid

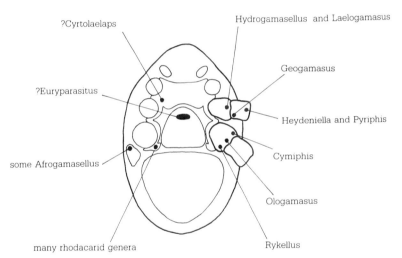

Figure 6.7 Schematic diagram of the ventral surface of a female rhodacarid mite with all legs removed except the basal segments of two rear legs, showing the "migration" of the female's copulatory pore from one site to another in different genera. In each genus the pore is connected to the spermatheca by a duct. Male chelicerae are used for intromission, and in at least some genera they are species-specific in form; hypodermic insemination is not known to occur. Neither the reasons for the movement of the female gonopore nor the intermediate stages in the process are known. The old observations of variant female crayfish with supernumerary gonopores on the legs that were connected to the oviduct (Bateson 1894) suggest that the evolutionary migrations may have occurred in relatively abrupt steps. (After Lee 1970.)

female sperm traps. As noted in the next section, it could also result from sexual selection by female choice. Whatever the explanation, female adaptations to males that are circumventing female structures put the females once again in the position of discriminating among males after copulation has occurred.

In summary, the observations presented in this section lend themselves to the view that postcopulatory processes within females are of sufficient selective importance to males that some males have evolved means of bypassing or manipulating these processes to their own reproductive advantage. Female evolutionary responses to such adaptations put the females of some groups back in a position to discriminate among males. The association of sperm-killing devices with the evolution of the new genitalic system of bedbugs and their relatives emphasizes the possible selective importance to females of controlling the postcopulatory events leading to fertilization.

Topology of female reproductive systems

One recurrent theme in the evolution of female genitalia is the production of apparently unnecessarily tortuous reproductive tracts morphology (Fig. 6.8). In some groups the female's behavior rather than her morphology per se produces situations that are analogously "perverse" from the male point of view: for example, the human cervix *withdraws* from the external opening of the vagina during sexual arousal (Masters and Johnson 1966). In the previous chapter I noted that it is probably advantageous to females to postpone fertilization until they are ready to oviposit. This may represent a point of conflict of interest between males and females; for males it is often advantageous to quickly fertilize as many eggs as possible if there is any chance that subsequent female behavior will result in any of those eggs being fertilized by other males. Here we may have both the "antagonistic" female morphology and the naturally selected advantage to the female of contravening males' interests, both of which are expected under the mechanical conflict of interest hypothesis but have been missing in many other cases.

Even in this case, however, sexual selection by female choice may have been involved, with the female using the male's ability to either traverse tortuous ducts himself or to induce her to transport

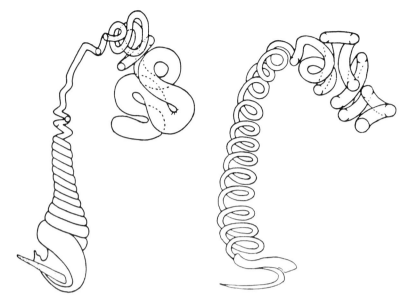

Figure 6.8 Spectacularly tortuous spermathecae and spermathecal ducts of two species of the hydraenid beetle genus *Meropathus*. The male's aedeagus has a long filiform extension that may penetrate at least partway up the ducts. Similar, highly coiled spermathecal ducts occur in a variety of animals, including spiders (Platnick and Shadab 1978, 1979) and hemipterans (Ahmad and Sheikh 1983). Possible explanations for such apparent obstacles to insemination include the mechanical conflict of interest hypothesis and sexual selection by female choice. (From Ordnish 1971, reprinted with permission of the author and editors of *Pacific insects Monographs*.)

his sperm through them as a screening mechanism allowing her to favor some males over others, the payoff to the female being the production of superior sons rather than avoidance of premature fertilization. In fact, both advantages could accrue simultaneously, and I see no way of separating the two hypotheses in all cases. In some groups, however, it is clear that the lengths of "overlong" female ducts do not correlate with the lengths of male intromittent organs: for example, the male hamster deposits sperm in the uterus, but the oviduct is extremely long and convoluted (Yanagimachi and Chang 1963); the aedeagus of the beetle *Altica* evidently does not enter the long, hairlike spermathecal duct (see Fig. 9.1). Under the conflict of interest hypothesis one might suppose that these female structures compensate for male abilities to *shoot* their sperm deeper into the female or for the motility of the sperm, rather than for physical properties of the male organs per se, but this seems an unlikely explanation for species specificity in the external morphology of the male organs. On the other hand, these female adaptations are explicable under the sexual selection hypothesis as parts of screening mechanisms; for instance, they might reward superior male ability to induce female transport of sperm along these tortuous passages.

The volume of female sperm storage organs relative to male ejaculate size is still another potentially important factor, since it could affect the probabilities of sperm precedence. Parker (1970) argued that males, since they produce cheaper gametes, must usually be strongly selected to fill up the female with more than enough sperm to fulfill all her anticipated needs for fertilization. In general the volumes of sperm transferred may be similar to the capacities of female storage organs, but there are at least two cases, honeybees (Kerr et al. 1962) and some odonates such as *Lestes* and *Sympetrum* (Waage, in press), in which female capacity is substantially greater than male ejaculate size. Honeybee males copulate only once (see Fig. 8.1), and their ejaculate size is perhaps limited by other demands such as feeding and flying, but odonate males often mate more than once in rapid succession (see, for instance, Thornhill and Alcock 1983; Waage, in press), so this explanation is not convincing. Waage presents evidence that males of these groups may pack the sperm of prior males into the blind ends of the females' bursae and spermathecae, displacing it from the vagina where it was placed during fertilization. Perhaps some oversize female storage organs are so large that it would be relatively costly for a male to attempt to fill them, and they may represent adaptations to test males' abilities to physically displace the sperm of other males.

One further puzzling female trait, the presence of multiple spermathecae (Fig. 6.9), may also be related, although critical data are lacking. For instance, some mecysmaucheniid spiders have up to 100 spermathecae (Forster and Platnick 1984), and the distantly related *Liphistius* spiders have up to 40 in each female (Murphy and Platnick 1981); there are three in each female *Drosophila* and many other flies (Imms 1964; Fowler 1973), and some other flies and the beetle *Blaps* have two (Imms 1964). In *Drosophila* the pair of smaller spermathecae fill first during copulation, but the sperm in the central ventral receptacle are usually used first to fertilize eggs; this

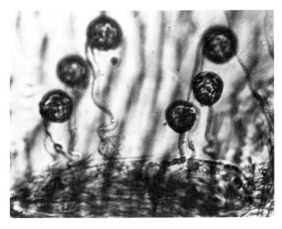
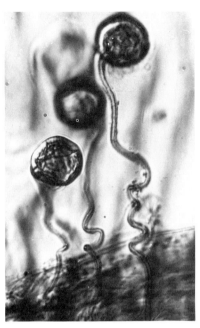
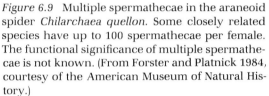

Figure 6.9 Multiple spermathecae in the araneoid spider *Chilarchaea quellon.* Some closely related species have up to 100 spermathecae per female. The functional significance of multiple spermathecae is not known. (From Forster and Platnick 1984, courtesy of the American Museum of Natural History.)

receptacle is often a long coiled tube whose size varies between species. In some species the two smaller spermathecae are gland-like and do not receive sperm (Fowler 1973). In the other groups little is known about how multiple spermathecae are used.

A second general trend in the evolution of female genitalia is the repeated derivation of separate copulatory openings, so that intromission and birth or oviposition occur through different openings in the female body (see Fig. 6.2). A conservative list of groups in which two separate female openings have originated includes spiders, probably more than once (Brignoli 1978; Opell 1983); mites, at least twice (Young 1968; Lee 1970); lepidopterans (Imms 1957); and flatworms (Hyman 1951b). Some insects (gerromorph bugs, Anderson 1981; some beetles, Imms 1957) have a variant morphology, with only a single external opening for both copulation and oviposition but a separate canal connecting the spermathica to the oviduct. The presumed sequence of evolutionary events in spiders was that a small outpocketing (spermatheca) of the oviduct to receive the sperm gradually became deeper and more sealed off from the oviduct and acquired first a groove along the oviduct and finally tubes of its own connecting it to both the exterior and the oviduct (Gertsch and Ennik 1983).

What advantages would accrue to females as they underwent such evolution? Clearly all stages represent increasing isolation of sperm from the oviduct, and increasingly indirect male access to fertilizable eggs. In the first stage (separation of spermatheca from oviduct) the advantage of such isolation could conceivably be an increased ability to maintain the stored sperm in special conditions favoring their survival. However, other stages, such as the separation of the insemination duct, do not seem explicable on these grounds, especially in groups like spiders, in which even in less derived groups the male intromittent organ usually reaches into the spermatheca, so there is no question of sperm having to survive in the insemination duct itself. A possibly important consequence of the separation of sperm duct from oviduct is that the female is able to control which sperm will fertilize her eggs. The "Loligo strategy" of flooding the oviduct with sperm (and perhaps even effecting fertilization before the female is ready to oviposit) is no longer possible for animals like spiders and moths in which the insemination canal is separate from the oviduct. Thus another factor responsible for the evolution of separate copulatory openings and associated ducts may have been the advantage to females of favoring fertilization by some males over others. Even if that was not the context in which insemination ducts originally evolved, the result is an increased probability that females have the ability to make such discriminations.

Finally, there are a few groups in which female morphology seems to be unusually "aggressive" or male-like, but the functional significance of the female structures is still unclear. In four groups of mites — glycyphagids, rosensteiniids, crypturoptids, and some *Trouessartia* (Alloptidae) — females have "copulatory tubes" or "external sperm ducts" (Santana 1976; Griffiths and Bocek 1977; OConnor and Reisen 1978; B. OConnor, personal communication). Copulation has never been observed in any of these groups, but in some the length of the female copulatory tube correlates inversely with the length of the male aedeagus (Santana 1976; OConnor and Reisen 1978; B. OConnor, personal communication). The female tubes are evidently rigid and simple, and in general vary only in length among different species. The tubes are apparently not intromittent organs, since males have aedeagi, and fertilization apparently occurs within the female. The tubes do not deposit eggs, as occurs in the ovipositors of the seahorses and their relatives for two reasons. First, the eggs are very large and could neither fit down the tube nor be lodged in any existing cavity in the male, and second, the oviducts empty to the exterior via a different pore. It seems very likely that in these groups the female investment in offspring is much greater than the male investment. The males of some species have species-specific structures to grasp and contact females (Santana 1976; B. OConnor, personal communication); in Chapter 11 I will argue that such devices function in courtship, so sex roles are apparently not inverted in courtship. Mechanical constraints may have made genitalic coupling difficult in some groups in which males use their abdomens in locomotion (B. OConnor, personal communication), but the functional significance of copulatory tubes is still a mystery.

Female genitalic "prehensors," which are complex and species-specific in form in some beetles of the small family Helodidae, were studied in detail by Nyholm (1969). In some genera the male inserts his aedeagus into the female and deposits a spermatophore, which the prehensor then grips. In these cases the prehensor may function as does the female bursal sclerite (signa) that is species-specific in many lepidopterans. In the beetle *Cyphon padi* the aedeagus is reduced, and the prehensor is inserted deep into the male's abdomen, where it seizes the spermatophore as it emerges from the male's ejaculatory duct. The spermatophore of this species is not large and has a long tail-like process that extends up the ejaculatory duct approximately three times the diameter of the sperm mass. The terminal processes of the prehensor are inserted *past* the sperm mass to grip the spermatophore near the tip of this tail. These beetles' intromittent prehensors thus differ from all other intromittent organs, male or female, in that they receive rather than donate gametes. Males of *Cyphon* are more active than females during courtship, and male genitalic processes anchor copulating pairs together. The possibility that some part of the spermatophore is nutritionally important for the female has not been explored, to my knowledge. The reason why females should evolve such unusual structures in these otherwise apparently unexceptional beetles is not clear, and the beetles clearly merit further study.

Summary

In some cases of genitalic evolution it is at present impossible to distinguish between sexual selection by female choice and sexual selection by direct male-male competition. The female choice hypothesis is more widely applicable, however; in many groups of animals with species-specific genitalia, direct male-male competition involving genitalic morphology is apparently physically impossible, but female choice could occur.

The repeated evolution of hypodermic insemination, of sperm storage organs that are displaced higher up in female reproductive tracts, and shifts in insemination sites on the female body are all possible consequences of selection on males to circumvent barriers in the female body that prevent immediate fertilization after each insemination. Several female characteristics, such as long, convoluted sperm ducts and separate external openings for copulation and oviposition may be female adaptations to prevent immediate fertilization. In cimicid bugs the clearly spermicidal properties of the female paragenital system that evolved in response to hypodermic insemination emphasize the possible importance of the female's ability to prevent immediate fertilization.

7 *Female discretion after genitalic contact*

This chapter addresses an apparent problem with the hypothesis of sexual selection by female choice that may have already occurred to the reader. If the hypothesis is true, females must be able to discriminate against some males that have already made genitalic contact with them. That is, copulation must not always result in the fertilization of all of the female's eggs by that particular male. But how can a female discriminate against a male once his genitalia are in contact with hers? By that time it seems too late to avoid fertilization by his sperm. I will show that not only is such discrimination feasible but the conditions necessary for it to occur are very widespread, and such discrimination has been documented. The clear conclusion is a sad one from a male's perspective: copulation does not always result in insemination, and insemination does not always result in fertilization. Females, because fertilization takes place within their bodies, generally have the last say in reproduction and can exercise what Thornhill (1983) has called "cryptic female choice."

Intromission

Commonly the female must respond appropriately to the male before successful intromission can occur, and often this happens after the male's genitalia first contact the female. In many groups such as insects, the female must open the tip of her abdomen or extend her ovipositor (Alexander 1959; Selander 1964; Wocjik 1969; West-Eberhard 1969; Pinto and Selander 1970; Eberhard 1981) or hold her abdomen up or to one side, as in some spiders (Bristowe 1958).

Females of many species can prevent intromission by such simple behavior as turning the abdomen away from the male, waving it from side to side, or simply moving away (Thornhill and Alcock 1983). This step is particularly dramatic in a species like the squid *Septioteuthis sepioidea,* in which the female sometimes detaches and discards a spermatophore that a male has placed on her (Moynihan and Rodaniche 1982). In those species in which the male taps or rubs the female with his genitalia before she allows intromission, females could discriminate among males on the basis of their genitalia. Especially in situations of high male density, as in mating swarms and around females emitting pheromones (for example, a moth, Doane 1968; a bee, Alcock and Buchman, in press), a female could select against a male by not allowing rapid intromission, thus increasing his chances of being displaced by another male.

Insemination

Copulation that fails, because of the female's behavior, to result in insemination has been documented in the mosquito *Aedes aegyptii* (Lea 1968; Gwadz et al. 1971), the fly *Musca domestica* (Adams and Hinz 1969), and the scorpionfly *Panorpa latipennis* (Thornhill 1980). In the scorpionfly insemination always occurred when a male offered food to his prospective mate but did not happen in 50 percent of the cases when a male tried to "rape" a female without offering her food; sperm counts in "rapist" males were not lower than in others. Females of the wasp *Polistes fuscatus* may also sometimes prevent insemination (Post and Jeanne 1983). In the midge *Culicoides melleus,* reduced amounts of sperm are sometimes transferred, apparently because of the female's behavior (Linley 1975). Male-female "incompatibility" sometimes reduces sperm transfer and/or use of sperm for fertilization in *Drosophila* (Fowler 1973), and some cross-specific pairs "pseudocopulate" without transferring any sperm (Markow 1981).

Neither the mechanisms used by the females to prevent insemination nor the factors that trigger their use is understood in any of these cases. It is interesting to note that female receptivity to insemination increases late in a female's life in *A. aegyptii* and *M. domestica.* Reduced female selectivity with age would be advantageous to avoid delaying oviposition, and this flexibility supports the argument being made here that female "discretion" is not arbitrary but is instead exercised in selectively advantageous ways.

Females can determine the length of copulation in many species (Thornhill and Alcock 1983, on insects), and if a female terminates copulation prematurely, fertilization may be less likely because less sperm or accessory fluid is transferred. This is probably true in many different groups. In the scorpionfly *Hylobittacus apicalis* (=*Bittacus apicalis*), for instance, the female determines the length of copulation; copulations of under 5 minutes result in the transfer of very few or no sperm; there is a positive linear correlation between numbers of sperm transferred and duration of copulation over the range of 5–20 minutes (Thornhill 1976). Gwynne (1982) found that females of the katydid *Conocephalus nigropleurum* sometimes pull away from the male after coupling but before a spermatophore is passed. McLain (1980) also showed that in virgin females of the bug *Nezara viridula* longer copulations reduce the level of sperm displacement resulting from subsequent matings. Longer copulations in the dung fly *Scatophaga* result in greater displacement of sperm already in the female (Parker, in Thornhill and Alcock 1983). And longer copulations in the salticid spider *Phidippus johnsoni* result in both higher likelihood of oviposition and lower likelihood of the female's remating (Jackson 1980). In species in which males defend females from the attentions of other males by remaining *in copulo* (Thornhill and Alcock 1983: in one insect species copulations last up to 79 days!), the defending male's ability to induce the female to allow longer copulations could increase his likelihood of obtaining fertilizations.

In many species copulations frequently fail to result in insemination, but the possibility that the female is responsible is untested. Particularly extensive data come from studies of moths and butterflies, in which females were checked for the presence of a spermatophore after they copulated. "Mated" females that lacked spermatophores were common: 9 percent in *Pectinophora gossypiella* (Ouye et al. 1965), 11 percent in *Grapholitha molesta* (Dustan 1964), 13 percent in *Danaus gilippus berenice* (Pliske 1973), 4 percent in *Lambdina fiscellaria lugubrosa* (Ostaff et al. 1974), 12 percent in *Choristoneura fumiferana* (Outram 1971), and 45 percent in *Diparopsis castanea* (Marks 1976 in Kirkendall, manuscript). Even spermatophore transfer does not always result in insemination; Etman and Hooper (1979) showed that 12 percent of the females of the moth *Spodoptera litura* that had received spermatophores nevertheless had no sperm in their spermathacae. When mating failures in moths were assayed by the inability of mated females to lay fertilized eggs, rates were 30–42 percent in *Apantesis* species (Bachelor and Habeck 1974), 34 percent in *Atteva punctella* (Taylor 1967), and about 30 percent in *Rhyaciona* and *Choristoneura* (Pointing and Campbell, in Taylor 1967). Mating failures were also common in *Dioryctria abietella* (Fatzinger and Asher 1971) and *Lymantria dispar* (Doane 1968) It is thus clear that mating failures are both widespread and intraspecifically common in many lepidopterans. Failures have generally been attributed to male dysfunction (Taylor 1967), but the alternative possibility, that females sometimes prevent insemination, has not been tested.

Sperm transport

As shown in Tables 1.1 – 1.3, males of animals with internal fertilization almost never deposit sperm directly on females' eggs. Instead, sperm are stored for a length of time, varying from hours to years, inside the female's reproductive tract before being used in fertilization. Usually the storage site is different from the fertilization site, and often the storage site is different from the insemination site. Thus sperm often must move from one place to another inside the female after copulation if fertilization is to occur (Fig. 7.1). In most groups the female rather than the sperm is responsible for this movement.

Walton (1960) argued that in mammals the movement of sperm from where they are deposited by the male (vagina in rabbit, cow, sheep, human, chimpanzee; uterus in horse, pig, dog, rat, mouse, hamster, guinea pig) toward the site where fertilization occurs (Fallopian tubes) is at least partly caused by contractions and movements of the female reproductive tract. This conclusion has been upheld by later studies (van Tienhoven 1968; Blandau 1969; 1973; Bishop 1971; Hunter 1975). It is also well established from artificial insemination studies in mammals that higher rates of conception result if semen is introduced higher up in the female tract — say in the uterus rather than the vagina (see, for example, Aamdal et al. 1978; Krzywinski and Jaczewski 1978) — so this transport is critical to the success or failure of a given copulation.

Sensory stimuli, including stimulation of the female genitalia, trigger contractions in her reproductive tract that generally result in sperm transport; this has been shown in cows (van Demark and Hays 1952, but see Blandau 1973), monkeys (Blandau 1969), rats (Adler 1969; Chester and Zucker 1970) and rabbits (Lambert and

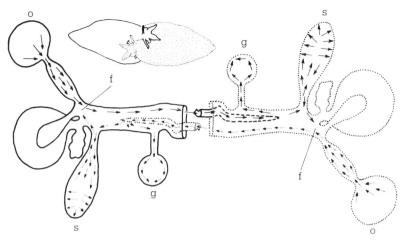

Figure 7.1 *Above*, mating pair of nudibranch molluscs, *Phyllaplysia taylori*. *Below*, complex routes of sperm migration during reciprocal insemination, traced using radioactively labeled sperm. Sperm move from the ovitestes (*o*) past the fertilization site (*f*) and out the penis into the other animal; from there they go either to the gametolytic gland (*g*) or the seminal receptacle (*s*). Some move from the receptacle to the partner's fertilization site (*f*). Sperm that are moved to the gametolytic gland are destroyed. Most movement of the sperm is due to the action of cilia lining the ducts rather than to their own propulsive force. Thus, inducing the partner to transport sperm is critical to achieving fertilization. Since many molluscs are capable both of self-fertilization and of sperm degradation at or near the site where sperm are introduced in copulation, female rejection of introduced sperm is an unusually feasible alternative. (After Beeman 1977.)

Tremblay 1978). In rabbits the number of artificially introduced sperm reaching the Fallopian tubes increased by a factor of four when artificial insemination was followed by copulation with a vasectomized male (Lambert and Tremblay 1978). In sheep a number of stimuli, including the presence of dogs, unfamiliar surroundings, and the act of artificial insemination, reduce the numbers of sperm reaching the uterus and Fallopian tubes; also, contractions of the uterus, and thus sperm transport, are inhibited by adrenalin (Robinson 1975). When artificially inseminated sheep were compared with naturally mated controls, sperm from artificial inseminations were found in the cervix in the same concentrations as sperm from natural matings and were found in greater numbers in the uterus, but occurred in much lower numbers in the Fallopian tubes (Allison 1975). This last result could be due to some inferiority of the artificially introduced sperm, which is the usual explanation given in artificial insemination studies or it could be due to differences in female transport of sperm. Levine (in Blandau 1969) made reciprocal crosses in genetically marked mice and showed that females selectively allowed some males to fertilize their eggs more often than others. Hunter (1975), speaking of mammals in general, noted that part of the transportation process is "undoubtedly" set in motion by the sexual stimulation associated with coitus and that it is promoted by oxytocin "released at coitus" (he also noted that chemical components of both rat and horse semen have a stimulating effect on smooth muscle — possibly an additional male adaptation to increase the probability of gamete transport).

Female-mediated transport of male gametes after copulation is also probable in insects. It has been documented by dissecting females at various times after copulation in the bug *Rhodnius* (Davey 1965) and in several lepidopterans (Norris 1932; Hewer 1934; Taylor 1967), and also by cutting abdominal nerves in female *Musca domestica* and showing that sperm did not reach the spermatheca (Degrugillier and Leopold 1972). Davey (1965) cited observations of contractions of the female duct in recently mated individuals in the orders Hemiptera, Lepidoptera, Neuroptera, Trichoptera, and Coleoptera. In lepidopterans and the beetle *Cyphon*, the female apparently also contracts her bursa, thus holding the spermatophore in place so that it empties sperm into the duct leading to the spermatheca (Proshold et al. 1975; see also Figs. 6.2 and 8.4). Chemical components of the semen of some insects induce contractions of the female ducts (Wigglesworth 1965). Chapman (1969) concluded that "the bulk of the evidence" suggests that females rather than the sperm themselves are responsible for sperm transport within female insects, and Walker (1980) gave additional references documenting the same conclusion.

Other animal groups are less well studied, but in some molluscs sperm are known to be moved in the female tract by muscular contractions (*Octopus*, Wells 1978) or ciliary movements (opisthobranchs, Beeman 1977; Fig. 7.1). Oldfield and colleagues (1972) observed rhythmic contractions of spermathecae and ducts in female *Aculus cornutus* mites; these probably result in sperm transport. In several groups the female morphology, such as ciliary linings in the regions of ducts where sperm are deposited during mating, where sperm are stored, and where fertilization occurs, suggests an active female role in sperm transport; this has been found in other molluscs (Rigby 1963), lizards (Cuellar 1966; Connor and Crews 1980), and salamanders (Boisseau and Joly 1975). The adaptive advantage to females of gamete transport being triggered by stimuli received during or just preceding copulation is obvious, and similar cues are probably used in many as yet unstudied species.

Sperm destruction, storage, and activation

Insemination does not automatically result in storage in species with internal fertilization. Digestion of sperm inside the female reproductive tract has been observed in mammals (Hunter 1975 on pigs and rats; Moghissi 1971 on humans), a fish (Philippi in Clark and Aronson 1951), an annellid worm (Picard 1980), stylommatophoran, pulmonate, and prosobanch molluscs (Rigby 1963; Duncan 1975; Webber 1977), turbellarian flatworms (Henley 1974), and insects (DeWilde 1964; Handlirsch, in Wigglesworth 1965; Proshold et al. 1975). Many molluscs have a "gametolytic gland," or bursa copulatrix (see Fig. 7.1), and studies in several groups "strongly suggest that its role is the digestion of excess sperm" (Duncan 1975, p. 337). Sperm are also evacuated via the vagina in some mammals (van Tienhoven 1968) and some insects (Englemann 1970; Fowler 1973; Woyke 1964). Male characteristics that inhibit any of these processes could be reproductively advantageous.

Females in many insects (probably most: see Wigglesworth 1965; Davey 1965) have spermathecal muscles that can decrease the volume of the spermatheca and thus probably function to move sperm onto eggs (Fig. 7.2). Villavaso (1975) showed that cutting the spermathecal muscle of the weevil *Anthonomus grandis* changed not only the female's ability to fertilize her eggs, but also the pattern of sperm precedence when two successive males were mated to her. Using radioactive markers to label the sperm of the first male, he assessed the relative contributions of the two males to the sperm retained in the female's spermatheca two days after the second copulation; in control females about 65 percent of the sperm were from the second male, while in operated females only about 20 percent were from the second male. Etman and Hooper (1979) showed that a second mating in the moth *Spodoptera litura* could cause the female to expel from her spermatheca the sperm received in a first mating. Thus the spermathecal muscle, which probably was originally designed to move sperm from a storage site to a

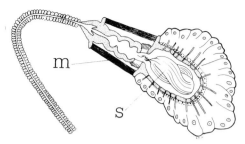

Figure 7.2 The syringelike spermatheca of the heteropteran insect *Gramphosoma*, with compressor muscles (*m*); when the muscles contract, sperm (*s*) are forced from the lumen down the spermathecal duct. The action of spermathecal muscles in ejecting sperm influences sperm precedence in some insects. Analogous mechanisms for moving sperm from storage to fertilization sites within the female (thus making the female potentially able to influence sperm precedence) exist in many other animal groups. (After Wigglesworth 1965.)

fertilization site, can also influence the pattern of sperm precedence from successive matings. Homologous and analogous structures undoubtedly exist in a wide array of animals that store sperm. It may be that the generally wide ranges of values found in studies of sperm precedence (see Lefevre and Jonsson 1962 on *Drosophila;* Jackson 1980 on spiders) are the result, at least in part, of females actively altering precedence by using spermathecal muscles or other mechanisms.

Although I know of no evidence regarding the control of accessory gland secretions into spermathecae or other sperm storage organs, it seems likely that such secretions too are often triggered to begin production or to release their contents by stimuli associated with copulation; to do so before copulation could waste energy and materials.

The females of a variety of animals, including mammals and insects (see Leopold and Degrugillier 1973), produce substances that "activate" sperm. Without these substances the sperm cannot carry out fertilization. Again, this is a female process that must take place before zygotes can be formed. I know of no evidence concerning how the release of these products is controlled. In some species, such as the fly *Musca domestica* (Leopold and Degrugillier 1973), the products may be released only when the mature egg descends the oviduct, usually long after copulation, but in other species production or release may occur closer to the time of copulation and be cued by stimuli associated with it.

Egg maturation and fetal nourishment

There are a number of examples of females modifying their production of eggs as a result of mechanical stimuli from copulation. Roth and Stay (1961) showed that stimulation of the female external genitalia in the roach *Diploptera punctata* caused oocytes to mature, that different artificial stimuli caused different numbers to mature, and that no artificial stimulus caused as many to develop as did stimulation by the male genitalia of *D. punctata* (Fig. 7.3). Gerber (1967) concluded that stimuli received during copulation affected both oocyte maturation and oviposition in the beetle *Tenebrio molitor*. Highnam (in Davey 1965) showed that stimuli from copulation triggered the release of hormones controlling vitellogenesis in the grasshopper *Schistocerca*. Copulation and male seminal fluid induce female *Drosophila* to lay larger eggs (Fowler 1973), and the same may be true in some crickets (Sakaluk and Cade 1983). Wigglesworth (1965) gave other examples in insects, and Davey (1965) stated that "it is clear" that mating often initiates or accelerates egg maturation in insects. El Said (1976) found that female *Amblyomma* ticks would not engorge and drop from their hosts until they had mated. In a number of species of mammals, stimuli from copulation are known to induce ovulation (Ewer 1973). Any male better able to induce these responses in females could have a reproductive advantage over others.

Still another process influencing the probability that a given male will sire offspring involves control of implantation of the fetus in the uterus of placental mammals and (probably) of analogous processes in a number of other groups, such as scorpions, sharks and rays, onychophorans, flies, mites, and cockroaches, in which the young are nourished inside the female's body. Mammals are apparently the only group in which the triggering of this process has been carefully studied. In some rodents the act of copulation triggers a hormonal cycle (luteal cycle) that results in preparation of the lining of the uterus for implantation. By inserting artificial penes into the vaginas of restrained female mice, Diamond (1970) found that the pattern of intromissions and thrusts influenced the likelihood that

Treatment	Age when treated (days)	Days after treatment	Oöcytes Mean ± S.E. (mm)	N
(1) Beeswax pellet inserted in bursa copulatrix	1 1	10 10	0·77 ± 0·04 1·30 ± 0·19	4 (−)* 3 (+)†
(2) Glass beads inserted in bursa copulatrix	1–2 1–2	8 8	0·72 ± 0·01 1·26 ± 0·16	19 (−) 5 (+)
(3) Dried spermatophores (expelled by mated females) moistened in Ringer's and inserted in the bursa copulatrix	1 1	11–12 11–12	0·69 ± 0·02 1·33 ± 0·12	9 (−) 5 (+)
(4) Beeswax plug inserted in uterine opening	1	10	0·68 ± 0·01	13 (−)
(5) Ovipositor cut off below the base	1 1	11 11	0·71 ± 0·01 1·63	11 (−) 1 (+)
(6) Ovipositor cut off at the base	<1 <1	7 7	0·72 ± 0·04 1·23	2 (−) 1 (+)
(7) Ovipositor cut off above the base	<1 2–4 2–4	8 11 11	0·69 ± 0·02 0·73 ± 0·02 1·14 ± 0·15	6 (−) 12 ( ) 4 (+)
(8) All valves of the ovipositor pinched with forceps	<1	6–10	0·71 ± 0·02	9 (−)
(9) Fleshy tissue below base of ovipositor pinched with forceps	6 6	11 11	0·67 ± 0·004 0·86 ± 0·01	7 (−) 3 (+)
(10) Fine forceps rubbed between valves and at base of ovipositor	3 3 3	7 11 11	0·71 ± 0·03 0·72 ± 0·02 0·88 ± 0·01	3 (−) 3 (−) 2 (+)
(11) Hot needle applied to outer region of bursa	<1 <1	6 6	0·64 ± 0·03 1·16 ± 0·04	2 (−) 3 (+)
(12) Hot needle inserted into bursa	5	6	0·73 ± 0·03	5 (−)
(13) Hot needle applied to ovipositor	5	6	0·73 ± 0·03	6 (−)
(14) Warm needle applied to ovipositor	5 5	10 17	0·69 ± 0·02 0·69 ± 0·01	6 (−) 5 (−)
(15) Drop of hot water (about 96 °C) applied to genital region	1 1	10 10	0·73 ± 0·02 0·92	4 (−) 1 (+)
(16) Male accessory glands (central lobes) crushed into a paste and applied to ovipositor	6 6	11 11	0·67 ± 0·02 0·96	7 (−) 1 (+)
(17) Viscous secretion of male accessory glands, taken from recently deposited spermatophores, applied to ovipositor	7 7	7 11	0·74 ± 0·03 0·73 ± 0·03	4 (−) 5 (−)

* (−) = No growth of oöcytes beyond that found in virgin females.
† (+) = Definite growth of oöcytes.

Figure 7.3 Manipulations performed on female *Diploptera punctata* cockroaches in attempts to induce oocyte development. Two important results were obtained: (1) tactile stimulation of the female genital area did induce oocyte maturation: and (2) none of the stimuli used were as effective as copulation with normal males of this species. (Reprinted with permission from *Journal of Insect Physiology*, vol. 7, Roth and Stay, "Oocyte development in *Diploptera punctata* [Eschscholtz] [Blattaria]," copyright 1961, Pergamon Press Ltd.)

the luteal cycle would be induced; again, the stimuli most nearly approximating those provided by conspecific males gave the greatest responses. McGill (1970) also claimed that the luteal cycle could be triggered more effectively by using an artificial penis shaped more nearly like that of the male (his data tended in that direction but were not statistically significant: $0.5 > p > 0.1$ with a chi square test). Chester and Zucker (1970) showed that induction of the luteal cycle in rats is influenced by patterns of intromission (presumably there was no difference in ejaculate size and composition in males performing different intromission patterns, but this possibility was not discussed). In contrast to these results, Kenney and colleagues (1977) failed to find differences in frequency of induction of the luteal cycle with different vaginal stimulations using artificial penes in seven other species of muroid rodents. Their sample sizes were quite small, however (generally less than 15), and, as they noted, their results may differ from those using laboratory species because of the wild species' reactions to handling.

Interestingly, the range of stimuli that induced the luteal cycle in rats was wider when the female was mated late in her estrous cycle; in nature, late sexual contact probably often correlates with lower chances of further copulations. Thus the females seem to "lower their standards" as time passes (as was shown above for flies and mosquitoes).

Oviposition rate

Another female reproductive process that is probably often triggered by copulation and/or insemination is oviposition, and some examples have already been mentioned. Experiments using castrated males have shown that stimuli from copulation itself or from the products of the male's accessory glands, rather than the presence of sperm in the female, increase subsequent oviposition rates in the grasshopper *Schistocerca gregaria*, the bug *Oncopeltus fasciatus* (Gordon and Loher 1968), and the beetle *Tenebrio molitor* (Gerber 1967). Heterospecific mating in *Oncopeltus* results in lower rates of oviposition (Chaplin 1973). Thornhill (1983) showed that in the scorpionfly *Harpobittacus nigriceps* the rate of oviposition immediately after copulation was positively related to both the size of the nuptial prey offered by the male and, when prey size was held constant, to the size of the male. McLain (1980) showed that once-mated females of the bug *Nezara viridula* were more likely to oviposit before remating if the first mating was of relatively long duration. In *Drosophila* either copulatory stimuli or male seminal fluid increases subsequent oviposition rates (Fowler 1973). Careful experiments have shown that in the honeybee *Apis mellifera* it is the full insertion of the male's endophallus (rather than partial insertion, external contact with the male, or deposition of sperm or accessory gland products) that releases oviposition behavior in the queen (Koeniger 1981; see Fig. 8.1). Again, the adaptive significance of such links is clear, for otherwise the female would lay unfertilized eggs, and similar responses probably will be found in many other groups that have not yet been studied in such detail.

Tendency to remate

In any species in which remating by the female results in some of her eggs being fertilized by the second male instead of the first (this is apparently very widespread — see next chapter), then a female can discriminate against a male by remating. Stated another way, the better a male is at inducing his mate to resist other males' attempts at copulation, the greater will be that male's reproductive success.

Probably stimuli resulting from copulation reduce females' receptivity to further matings in a very large number of groups. An undoubtedly incomplete list of animals in which this has been demonstrated experimentally includes salamanders (Halliday 1977), *Rattus norvegicus* (Bermant and Westbrook, in Doty 1974), butterflies (Labine, in Taylor 1967; Boggs 1981; Rutowski, in Boggs 1981), mosquitoes (Craig 1967), houseflies (Riemann et al. 1967), *Drosophila* (Manning in Fowler 1973), other flies (Leahy 1967), the honeybee *Apis mellifera* (Woyke 1964), several grasshoppers and a roach (Englemann 1970), and a tettigoniid katydid (Morris et al., in Cade 1979). Dewsbury (1981) cites additional evidence for several vertebrates. Disinclination to remate is probably a generally advantageous trait for females because mating is dangerous (Daly 1981) and often costly in terms of time and energy (Richards 1927b; Parker 1970). It is likely that copulation reduces female receptivity in many other groups.

Summary

In summary, it is clear that before copulation can result in fertilization and the production of offspring, a number of different female processes must occur, including facilitation of intromission and insemination; inhibition of sperm destruction; transport, storage, and activation of sperm; egg maturation and fetal nourishment; induction of oviposition; and inhibition of the tendency to remate. Natural selection on females to avoid wasted energy and materials favors the use of cues associated with copulation to trigger many of these processes.

This female responsiveness to male cues must predispose the male structures that stimulate females in sexual contexts to undergo bouts of runaway evolution. The male adaptations noted in Chapter 6, which circumvent female structures that impede their sperm from having direct access to fertilizable eggs testify to the fact that the postcopulatory processes in females are important to the reproductive interests of males.

It is interesting to note that most of the previous hypotheses explaining genitalic evolution did not take into account the possibility of female discretion (even genitalic recognition as originally proposed, was thought to occur in the male, Jeannel 1941). In fact several of the experimental tests of the hypotheses failed to give conclusive evidence because they did not deal with female processes following copulation (see Chapter 2). It is tempting to attribute this historically persistent oversight to the pervasiveness of a male viewpoint in biology. In this respect, perhaps the sexual selection by female choice hypothesis is a product of its times.

8 *Remating by females*

Another pair of conditions is necessary to the hypothesis of sexual selection by female choice: first, females must make genitalic contact with more than one male of their own species relatively often; and second, remating must result in at least some fertilizations being lost by the first male and gained by the next one. The second condition, which involves sperm precedence within the female, seems to be met in all species of arthropods that have been investigated (Parker 1970; Walker 1980; Gywnne 1983; Austad 1982) and is probably also true for other groups (see Hanken and Sherman 1981, on ground squirrels). I will not discuss this condition further here.

Courtship and copulation are often dangerous as well as costly in both time and energy (Richards 1927b; Daly 1978), so remating by females seems paradoxical. A number of possible reasons have been proposed — Walker 1980 lists eleven explanations in four major categories. It is worth noting that once sexual selection by female choice began to act on a species' genitalia, it would generate selection on females to perpetuate the tendency to remate, since by mating with several males a female would enhance her chances of encountering an especially capable one. Thus sexual selection by female choice could help to maintain the necessary conditions for its own existence.

This chapter is devoted to examining evidence on female remating frequencies. It may seem unlikely that multiple genitalic contacts with males could be common enough to account for widespread occurrence of rapid and divergent genitalic evolution, and that the rarity of such contact constitutes reason to reject the female choice hypothesis. There is a widespread assumption that females of many species mate only once (see, for example Wigglesworth 1965 on insects), but it turns out that there is little field evidence to substantiate this. Instead, the existing data suggest that remating by females is widespread. Before reviewing that evidence, I will examine four types of cases in which female remating seems especially unlikely.

Strict behavioral monogamy induced by males

In *Aedes aegyptii* mosquitoes and *Musca domestica* flies, the male deposits substances in the female during insemination that render her almost completely refractory to further matings for the rest of her life* (Craig 1967; Riemann et al. 1967). There is enough repressor substance in a single ejaculate of *A. aegyptii* to repress sixty-four females (Craig 1967). Surprisingly, however, this effect does not preclude the possibility of sexual selection on male genitalia in these groups, both of which have species-specific male genitalia. This is because in these species females are known to copulate repeatedly (more than a hundred times in *A. aegyptii* kept in the lab, Gwadz et al. 1971) before finally allowing a male to inseminate them. In the wild, *M. domestica* females probably do copulate repeatedly before they are inseminated, since both males and females are attracted to common feeding sites, and females thus have ample opportunity to encounter males. In fact the females' persistent prevention of insemination is not easy to explain on any grounds other than as a mechanism to discriminate among males.

In both species the female's tendency to permit insemination increases over time but is not strictly correlated with age; that is, not all females allow insemination at some fixed age; instead, gradually increasing numbers permit insemination as time passes (Adams and Hinz 1969; Lea 1968; Gwadz et al. 1971). Even supposedly receptive virgin females of *M. domestica* fail to receive sperm in the spermatheca in 30–35 percent of copulations in captivity (Leopold and Degrugillier 1973). As noted in Chapter 7, increased receptivity with

---

* If the female remated within a short period of time (on the order of hours) in the laboratory, a second insemination occurred in 22 percent of 90 cases in *Musca domestica* (Degrugillier and Leopold 1972); Baldwin and Bryant (1981) found that about 4 percent of the females of this species remated when the second male to which they were exposed was larger than the first.

age is probably adaptive, since in nature the older a female is the first time she contacts a male, the more probable it is that males occur at low densities; the female cannot afford to be as discriminating because of the greater danger of going unfertilized. The variation in the age at which a female finally accepts a male is generally attributed to asynchrony among females, but it could also be due to differences in the males' abilities to stimulate females to allow insemination. Certainly there would be strong selection on males to stimulate the female to allow insemination, since success would mean siring the female's entire reproductive output.

Copulatory plugs

In a number of groups the male deposits a "plug" of one sort or another that is thought to exclude other males from intromitting; this is known in many genera of Lepidoptera (Taylor 1967; Ehrlich and Ehrlich 1978) and in the mosquitoes *Anopheles* and *Aedes* (Chapman 1969), the bug *Rhodnius* (Davey 1960), the firefly *Pteropteryx* (Wing 1982), some dyticid beetles (Balduf, in Sivinski 1980), the crayfish *Cambarus* (Andrews 1904), various nematodes (Hope 1974), an acanthocephalan worm (Abele and Gilchrist 1977), several different spiders (Levi 1959; Jackson 1980; Robinson 1982; G. B. Edwards and H. W. Levi, personal communication), some ticks (Oliver et al. 1974), some rodents (Voss 1979), the stingless bee *Melipona quadrifasciata* (Kerr et al. 1962), and the honeybee *Apis mellifera* (Michener 1974). In the bees the entire genitalic apparatus of the male is detached and remains attached to the female genitalia, as shown in Fig. 8.1 (see Parker [1970] for other examples of plugs in insects). In at least some of these groups (lepidopterans, rodents, *Pteropteryx*, *Apis*, *Cambarus*, the spiders) the male genitalia are species-specific (respectively, Klots 1970; Prasad 1974; Ballantyne and McLean 1970; C. Michener personal communication; Pennak 1978; Kaston 1948).

Nevertheless, in those species for which there are relatively complete observations (sample sizes of 30 or more) on whether females remate, the copulatory plugs, although they may reduce the frequency of female remating, apparently do not preclude subsequent copulations. As Taylor (1967) pointed out, more than one plug (presumably indicating more than one copulation) has been found in females of the lepidopterans *Parnassius*, *Argynnis*, *Euphydryas*,

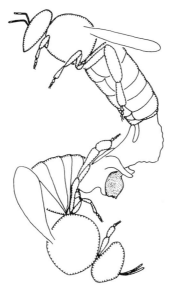

Figure 8.1 A male honeybee *(Apis mellifera)* hanging below a queen with which he copulates during her nuptial flight. In this species copulation is suicidal for the male; the posterior portion of his genitalia explodes, leaving the tip of his abdomen attached to the female's genitalia when he drops free. The male's sacrifice is not sufficient to prevent further matings by the female, however. The queen's next mate removes the plus (stippled in the drawing) before inserting his own genitalia. Koeniger (1981a) has shown that stimuli resulting from the insertion of the male's genitalia induce oviposition in the queen. (After Koeniger 1983.)

Polyommatus, Acraea, and *Bombyx.* Ehrlich and Ehrlich (1978) failed to find clear differences between the remating rates (as deduced from spermatophore counts in females) of 49 species that made copulatory plugs as compared with 22 which did not; they concluded that their results suggest "that plugging has no significant effect on multiple mating" (p. 680). Female *Cambarus* crayfish will mate with several males (Andrews 1904). Female ticks with plugs have been found with up to five spermatophores inside them, and a single spermatheca of the spider *Latrodectus* may contain up to five different embolus tips (Abalos 1968; also Fig. 8.2). In 52 of 74 female *Phidippus johnsoni* spiders, plugs in the genitalia were displaced as a result of a second copulation (Jackson 1980). In acanthocephalan worm females the plug falls out as soon as two days after copulation, and eggs are sometimes shed for more than a hundred

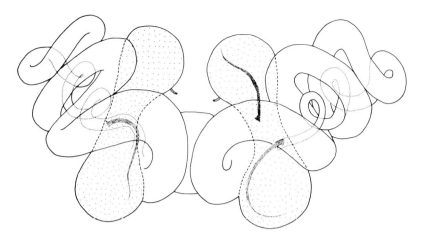

Figure 8.2 The spermathecae (light stippling) and the tortuous sperm ducts of a black widow spider, *Latrodectus,* with three broken tips (dark stippling) of the male intromittent organ (embolus) inside. The embolus has a clear dehiscence zone where the tip breaks, suggesting that the breakage is advantageous to the male. It is clear that the broken tips do not exclude subsequent males: up to five tips have been found in a single spermatheca. (Modified from Abalos 1968.)

days (Abele and Gilchrist 1977). Rodent females sometimes turn and gnaw the plug out of their vagina soon after mating, and in many species the lining of the vagina is shed so that the plug is discarded (R. Voss 1979, and personal communication). *Apis mellifera* queens usually mate several times on each nuptial flight despite the short durations of the flights and the presence of previous males' genitalia (Michener 1974); each male removes the genitalia of the previous mate, as shown in Fig. 8.1 (Koeniger 1983). Thus, as far as I have been able to ascertain, deposition of plugs does not preclude remating by the female in any group with species-specific genitalia. Nor, of course, does it rule out previous genitalic contact which did not result in insemination and plug deposition.

If deposition of copulatory plugs in a given group were invariably associated with copulation and if it consistently eliminated the possibility of further inseminations, the female choice hypothesis would predict lack of rapid divergence in the evolution of genitalic structures. As noted, I have found no evidence that such plugs are permanent, so the prediction remains untested. On balance, evidence from copulatory plugs does not rule out the possibility of sexual selection by female choice on male genitalia.

Antiaphrodisiacs

Males of some species deposit in or on their mates substances that appear to function as a sort of pheromonal or psychological mating plug by inhibiting other males from responding to the female. This has been noted in the beetle *Tenebrio molitor* (Happ 1969), the moth *Pseudaletia unipuncta* (Hirai et al. 1978), and perhaps the butterfly *Heliconius erato* (Gilbert 1976; see also Thornhill and Alcock 1983). It turns out, however, that these "antiaphrodisiacs" do not deter females from remating — in fact *T. molitor* females are known to mate up to six times in "several hours" (Happ 1969), and almost half of the mated *P. unipuncta* females captured in the field had more than one spermatophore (Callahan and Chapin 1960). The biological significance of the antiaphrodisiac effect in *Heliconius* is yet to be demonstrated, since according to Gilbert (1976) the odor, after being transferred to the females, was released only when females were handled, and only tests of male reactions to females that were being handled were reported. *Heliconius erato* females in nature do certainly remate; the spermatophore counts in Table 8.1 show 8 percent remating. The conclusion here is thus the same as for plugs: male deterrence to remating may occur, but male-enforced monogamy has not yet been documented in any group in which genitalia are species-specific.

Short-lived females

Remating by females also seems particularly unlikely among animals whose adult lives are very short, as in many species in the order Ephemeroptera (mayflies), some of which live as adults for as little as 90 minutes (Edmunds et al. 1976). Yet in most groups the male genitalia are species-specific (Edmunds et al. 1976). There are apparently no careful studies of whether or not female mayflies remate; it is generally assumed, without evidence, that they do not (see Spieth 1940). As shown by the observation of a *Paraleptophlebia debilis* female flying into a swarm, copulating, then flying on into another swarm and copulating again (Edmunds et al. 1976), remating can occur within a very short time; mayfly males generally transfer sperm very rapidly (Edmunds et al. 1976).

Monogamy in termites

Strict monogamy is well documented in termites. Before copulating, the founding male and female actually seal themselves into a chamber, thus initiating the new nest, which they will never leave (Wilson 1971). The prediction of the female choice hypothesis is that termite genitalia will be simple and uniform, and this is clearly upheld (Roonwal 1970; Fig. 8.3). This confirmation is especially dramatic because termites are derived from ancestors that resemble roaches or mantids (Imms 1957), the contemporary groups of which have complex genitalia. The genitalic simplicity of termites is probably secondarily derived.

Determining remating frequencies

Determination of remating frequency in nature by following individual females for their entire reproductive lives is probably not feasible for most species. However, Wing (1982) pointed out a much simpler technique for testing whether females remate after being inseminated. If one can determine unequivocally that a female is receptive to mating before she actually copulates, dissection to check for sperm in her reproductive tract will reveal whether she has been inseminated previously. If one can be sure that the sites where females are sampled are the only places where mating occurs and that all copulations result in inseminations, one can test whether females remate.

The existing data of this kind suggest that the females studied do remate. Female crickets that are attracted to male calling songs, in all species for which data exist, are often nonvirgins (Ulagaraj 1975; Forest 1983, on two species of the mole crickets *Scapteriscus;* Walker 1979 and personal communication, on *Anurogryllus arboreus;* Cade 1979a on *Gryllus integer*). Nonvirgin females of the dynastine beetle *Podischnus agenor* also come to the tunnels of males emitting attractant pheromones (personal observation). Interpretation of the cricket data is complicated by the fact that the

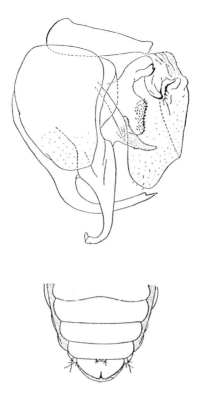

Figure 8.3 Above, the complex male genitalia of the mantis *Polyspilota
aeruginosa* and, *below,* the simple male genitalia of the termite *Odonto-
termes obesus.* Termites are probably descended from mantidlike or
roachlike ancestors; both mantids and roaches have relatively complex and
often species-specific male genitalia. Thus termite genitalia have probably
evolved toward reduced complexity. This simple genitalic structure is predicted by the female choice hypothesis because most female termites are
strictly monandrous. (From Beier 1970 and Roonwal 1970.)

female sometimes uses male songs not to find mates but to locate
suitable habitats; for instance, some *Scapteriscus* females land near
males but do not mate (Forrest 1983). Even so, it appears that in all of
the species just named some of the nonvirgin females mate with the
calling male. Studies of this sort on female mayflies flying into mating swarms would be especially interesting as a test of the sexual
selection hypothesis.

Many studies have been made in captivity, showing that remating
does or does not occur in a given species. Neither conclusion is
convincing, however. Such studies often inundate females with

males, which may result in unnaturally high frequencies of remating (on *Drosophila*, Gromko and Pyle 1978; on moths, Perez and Long 1964; Raulston 1975). Also, by definition, the observations are not made under natural conditions, and critical stimuli needed to induce females to remate may be lacking, resulting in unnaturally low remating frequencies. For instance, statements that females of the moth *Choristoneura* cease to attract males after mating (see Greenbank 1963, in Outram 1971) seem to give unreliable predictions of remating frequencies in the field, since spermatophore counts in females captured in nature show that females often remate (Outram 1971).

There are numerous scattered observations of females remating under natural conditions. But these cannot be taken as indicative of general tendencies either, since observers are seldom able to follow a female through her entire adult life and thus cannot affirm that she does not remate (Page and Metcalf 1982). Observations of females remating can, on the other hand, be made relatively easily, so reports from field studies are undoubtedly biased toward remating.

Many studies also show that males compete for and copulate with newly mature females and that females leave the emergence sites after a single copulation. This is sometimes assumed to indicate that females mate only once (Alcock et al. 1976). However, at least some of these species can and do also mate at other sites. For instance, a female *Trigonopsis cameronii* wasp that emerged and left the nest site without copulating was nevertheless fertilized (Eberhard 1974; male genitalia in this genus are species-specific, Vardy 1978). Some males of the bee *Centris pallida* patrol areas where females never emerge and presumably mate there at least occasionally (Alcock 1979; again, male genitalia are species-specific, J. Alcock, personal communication). The female lobster *Homarus americanus* generally mates only just after she moults, and she is usually guarded by a male during this period, but nevertheless females are sometimes mated more than once (Nelson and Hedgecock 1977). Females of the mosquitoes *Opifex fuscus* and *Deinocerites cancer* are mated as they emerge from their pupal skins but may remate occasionally (especially *O. fuscus*); males attempt to sequester the females, suggesting that remating has been frequent enough to be a factor in the evolution of the males' behavior (Provost and Haeger 1967). The need to be cautious in deciding whether or not females mate with more than one male is emphasized by the discovery of common polyandry in relatively well-studied bird species that were long thought to be monandrous (Sherman 1981 and references), and in the deermouse *Peromyscus maniculatus* (Birdsall and Nash 1973), which was also thought to be monandrous (Kleiman 1977).

Genetic analysis of offspring is another method of estimating frequencies of rematings by females in the field, but it is more arduous and somewhat uncertain. Such "paternity exclusion" analyses generally depend on electrophoretic studies and give only ranges of multipaternity, since genetically similar males cannot be distinguished directly and must be estimated using probabilistic formulas (Merritt and Wu 1975). The results are also underestimates, at best, of frequencies of matings with different males, since copulations that do not result in insemination or fertilization are not taken into account. Multiple paternity has been found in the deermouse *Peromyscus maniculatus* (Birdsall and Nash 1973), the ground squirrel *Spermophilus beldingi* (Hanken and Sherman 1981), the lobster *Homarus americanus* (Nelson and Hedgecock 1977), the snail *Cepaea nemoralis* (Murray 1964), the spider *Prolinyphia marginata* (Martynuik and Jaenike 1982), the wasp *Polistes metricus* (Metcalf and Whitt 1977), and in the flies *Drosophila pseudoobscura* (Cobbs 1977; Levine et al. 1980), *D. melanogaster* (Milkman and Zeitler 1974), and *Dacus oleae* (Zouros and Krimbas 1970). Monogamy, but with some serial polygamy, was found by Foltz (1981) in females of the mouse *Peromyscus polionotus*.

Another technique involves comparing the quantity of sperm in the female's reproductive tract with the amount ejaculated by a single male. This technique can give definitive demonstrations of multiple mating if a single male cannot supply the quantity found in a single female (on *Apis mellifera*, see Kerr et al. 1962), but it can give only weak indications of monandry (Page and Metcalf 1982). *Melipona quadrifasciata* is a widely cited case (Crozier 1977; Page and Metcalf 1982) of supposed monandry in a group with species-specific genitalia (C. D. Michener, personal communication). Kerr and coworkers (1962) dissected two females, and the numbers of sperm they found in each corresponded closely to the amount of sperm contained in a single male, but not to the amount ejaculated; the same authors presented data from the honeybee suggesting that males ejaculate less than their total store of sperm. Their conclusion, that females mate only once, seems unjustified, even aside from the tiny sample size, since (1) ejaculate amounts were not determined; (2) the possibility was not excluded that some copulations resulted in reduced or negligible amounts of sperm being transferred, or that some sperm were later discarded; and (3) matings within nests involving gravid queens have been documented in Meliponinae (Michener 1974) and might not have been noticed by Kerr and colleagues in their observation hives.

The best field data on remating frequencies

Frequencies of matings by females in the field are best known in Lepidoptera because the male deposits a durable spermatophore in the bursa of the female (Fig. 8.4; see also Fig. 6.2). In many species part of the spermatophore (the "neck") remains intact in the bursa

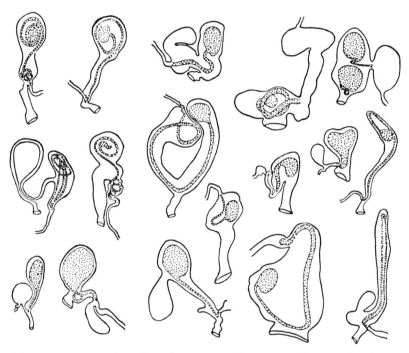

Figure 8.4 A sampler of moth spermatophores (stippled) as they lie in female bursae. All are from different genera except congeneric pairs at the upper right and upper left. The sperm leave the narrow tip of the spermatophore and are transported along the "seminal duct" to the spermatheca (not shown, see Fig. 6.2). Each spermatophore is carefully positioned so that its opening lies opposite that of the duct. By counting spermatophores or their remains in field-captured females, one can estimate rates of female remating under natural conditions; the bursae at top right and bottom right each have two intact spermatophores. (After Williams 1941.)

for the female's entire life. Thus, by counting the number of necks in the bursae of field-captured females, one can deduce how often females have mated. Similar counts are apparently feasible in other groups, such as some mites (Griffiths and Boczek 1977), some spiders (Abalos 1968; Levi 1975), and tettigoniid orthopterans (Gwynne, manuscript). Only in the black widow, *Latrodectus,* are data available; in three species (50 females each), the minimum estimates of the average number of matings per female were 1.14, 1.20, and 1.24, and the maximum numbers per female were 3, 3, and 2 (Abalos and Baez 1966).

Table 8.1, based in large part on an unpublished compilation made by Larry Kirkendall, summarizes the data on spermatophore counts in Lepidoptera in the field and shows that at least occasional remating is very widespread. These remating rates are probably underestimates for several reasons. Females of some species are known to lose or completely destroy spermatophores (Burns 1968, Taylor 1967, Ehrlich and Ehrlich 1978, Boggs 1981). It is also common in Lepidoptera for copulations to fail to transfer a spermatophore (the frequencies range from 4 percent to 45 percent; see Chapter 7). Under some conditions — probably rare in nature, however — males of the fruit moth *Grapholitha* can transfer sperm without passing a spermatophore (George and Howard 1968). Finally, since older females had mated more times than younger ones in the species in which this was checked (Dustan 1964; Burns 1968; Pease 1968; Outram 1971; Pliske 1973; Ehrlich and Ehrlich 1978), one can assume that most samples contain some young females that would have mated additional times if they had not been captured.

One must also keep in mind that dramatic differences in a species' remating frequencies have been documented at different times of year, in different years, and at different sites. For instance, 7 percent of the young females of the Oriental fruit moth *Grapholitha molesta* had remated in the spring generation of 1962, while 19 percent had remated in the fall generation of that year, and 22 percent had remated in the spring generation of the following year (Dustan 1964). Differences between broods have also been documented in the codling moth *Carpocapsa pomonella* (Gehring and Madsen 1963) and the monarch butterfly *Daneus* (Williams et al. 1942, in Pliske 1973). Gehring and Madsen (1963) showed variation of almost 400 percent between different sites in females of the same generation of the codling moth *Carpocapsa pomonella.* Burns (1968) also showed that mating frequencies of the females of some skipper butterflies correlated inversely with population density (neither

Table 8.1 Spermatophore counts in field-captured females of 94 lepidopteran species. Only those species in which at least 10 females have been examined are included; for additional data, see Ehrlich and Ehrlich 1978. These data give conservative estimates of the number of times females mate under natural conditions and show that multiple mating by females is widespread.

Species	Ratio of females with more than one spermatophore to total females with spermatophores	Average no. spermatophores in females with at least one spermatophore	Maximum no. spermatophores	N	References
MOTHS					
Gelechiidae					
Pectinophora gossypiella[b]	.18[a]	1.22[a]	6	2,570	Graham et al. 1965
Noctuidae					
Diparopsis castanea[b]	.47	1.78	4	128	Marks 1976
Heliothis zea[b]	.41[a]	1.56[a]	5	1,295	Pease 1968
H. virescens[b]	.71[a]	2.61[a]	—	238	Raulston et al. 1975
Peridroma saucia	.56[a]	2.19[a]	5	239	Callahan and Chapin 1960
Pseudaletia unipuncta	.58[a]	1.84[a]	5	645	Callahan and Chapin 1960
Arctiidae					
Utetheisa ornatrix[b]	.85	3.84	11	88	Connor et al. 1980; Pease 1968
Pyralidae					
Diatraea saccharalis[b]	.02	1.02	5	674	Perez and Long 1964
Acrobasis vaccinii	.69	2.55	9	3,604	Tomlinson 1966
Cossidae					
Prionoxystus robiniae[b]	.01	1.00	—	>200	Solomon and Neel 1973

Olethreutidae					
Grapholitha molesta[b]	.18–.25	1.30		77	Dustan 1964
Carpocapsa pomonella	.05–.30	—		—	Gehring and Madsen 1963
Tortricidae					
Choristoneura fumiferana[b]	—	≫ 1.0[c]		—	Outram 1971
Lymantriidae					
Lymantria dispar[b]	~0	~1.0		—	Doane 1968
BUTTERFLIES					
Papilionidae					
Battus philenor	.48[d]	1.73[d]	5	33	Burns 1968
Papilio aegeus	—	1.33[d]	2	10	Ehrlich and Ehrlich 1978
P. glaucus	.42	1.50	5	555	Burns 1968; Makielski 1972; Pliske 1973
P. palamedes	.40[d]	1.47[d]	3	32	Pliske 1973
P. rutulus	—	1.47[d]	3	18	Ehrlich and Ehrlich 1978
P. troilus	.20	1.24	4	358	Pliske 1973
P. zelicaon	.12[a]	1.16[a]	3	112	Shields 1967
Parides anchises	—	1.29[d]	3	17	Ehrlich and Ehrlich 1978
Pieridae					
Ascia monuste	.18	1.20	3	99	Nielsen 1961; Pliske 1973
Anthocaris cethura	0[a,d]	1.00[a,d]	1	26	Shields 1967
A. sara	.18[d]	1.11[d]	2	28	Shields 1967
Colias alexandra	—	1.24[d]	2	17	Ehrlich and Ehrlich 1978
C. philodice	.19	1.21	3	106	Stern and Smith 1960; Ehrlich and Ehrlich 1978

Table 8.1 (continued)

Species	Ratio of females with more than one spermatophore to total females with spermatophores	Average no. spermatophores in females with at least one spermatophore	Maximum no. spermatophores	N	References
Euchloe ausonides	.56[d]	1.74[d]	3	43	Ehrlich and Ehrlich 1978
Eurema lisa	.32[d]	1.35[d]	3	31	Rutowski 1978
Pieris napi complex	.33[a,d]	1.42[a,d]	3	27	Ehrlich and Ehrlich 1978
P. protodice	.03[d]	1.03[d]	2	32	Shields 1967; Enrlich and Ehrlich 1978
P. rapae	.74[d]	2.66[d]	5	49	Burns 1968
Satyridae					
Cercyonis pegala	.04[d]	1.04[d]	2	28	Burns 1968
Coeconympha tullia	.06[a]	1.07[a]	2	102	Shields 1967; Ehrlich and Ehrlich 1978
Erebia epipsodea	.19	1.21	3	197	Ehrlich and Ehrlich 1978
Euptychia hermes	.09	1.09	2	129	Pliske 1973; Ehrlich and Ehrlich 1978
E. hesione	—	0.91[e,d]	1	13	Ehrlich and Ehrlich 1978
E. libye	—	1.15[d]	2	16	Ehrlich and Ehrlich 1978
E. palladia	—	0.80[d,e]	1	12	Ehrlich and Ehrlich 1978
E. penelope	0[d]	1.00[d]	1	18	Ehrlich and Ehrlich 1978
E. renata	.18[d]	1.18[d]	2	22	Ehrlich and Ehrlich 1978
Neominois ridingsii	.03	1.03	2	63	Scott 1973b
Oeneis chryxus	.19	1.23	3	50	Ehrlich and Ehrlich 1978
O. taygete	—	1.70[d]	3	11	Ehrlich and Ehrlich 1978
O. uhleri	—	1.13[d]	2	16	Ehrlich and Ehrlich 1978

Nymphalidae					
Anartia amanthea	.15	1.16	3	119	Pliske 1973; Ehrlich and Ehrlich 1978
Biblis hyperia	.59[d]	1.95[d]	7	41	Pliske 1973
Catonephile numilia	—	1.60[d]	3	11	Ehrlich and Ehrlich 1978
Chlosyne acastus	.07[d]	1.11[d]	3	28	Shields 1967; Ehrlich and Ehrlich 1978
C. palla	.10[d]	1.10[d]	2	21	Ehrlich and Ehrlich 1978
Euphydryas anicia	—	1.22[d]	2	24	Ehrlich and Ehrlich 1978
E. editha	.31	1.32	2	86	Ehrlich 1965; Labine 1966; Ehrlich and Ehrlich 1978
Euptoieta claudia	0[d,a]	1.00[d,a]	1	10	Ehrlich and Ehrlich 1978
Poladryas minuta	.03[a]	1.03[a]	2	51	Scott 1974b
P. pola	0[a,d]	1.00[a,d]	1	11	Shields 1967
Precis lavinia	.45	1.50	3	102	Pliske 1973
Pteronymia artena	—	1.50[d]	2	10	Ehrlich and Ehrlich 1978
Speyeria callippe	.04	1.04	2	55	Shields 1967
S. cybele	.05	1.05	2	68	Burns 1968
Tithorea harmonia	—	1.46[d]	2	11	Ehrlich and Ehrlich 1978
Heliconiidae					
Agraulis vanillae	.16	1.16	2	102	Pliske 1973
Heliconius aliphera	—	1.69[d]	3	13	Ehrlich and Ehrlich 1978
H. erato	.08	1.09[e]	3	120	Pliske 1973; Gilbert 1976; Ehrlich and Ehrlich 1978
H. ethilla	.04[d]	1.04[d]	2	28	Ehrlich and Ehrlich 1978
Phyciodes campestris	.10[d]	1.10[d]	2	22	Ehrlich and Ehrlich 1978

Table 8.1 (continued)

Species	Ratio of females with more than one spermatophore to total females with spermatophores	Average no. spermatophores in females with at least one spermatophore	Maximum no. spermatophores	N	References
Daneidae					
Daneus gilippus (Texas)	.65	2.63	10	50	Burns 1968
D. g. berenice	.77	4.02	15	193	Pliske 1973
D. g. xanthippus	.57	2.35	8	70	Pliske 1973
D. plexippus	.52	2.23	8	91	Pliske 1973
Riodinidae					
Nymphidium cachrus	—	3.13[d]	7	15	Ehrlich and Ehrlich 1978
Lycaenidae					
Agriades glandon	.05[d]	0.83[e,d]	2	27	Ehrlich and Ehrlich 1978
Celastrina argiolus	0[d]	1.00[d]	1	12	Shields 1967
Deudorix antalus	—	.10[e,d]	1	10	Ehrlich and Ehrlich 1978
Everes amyntula	—	1.00[d,e]	2	12	Ehrlich and Ehrlich 1978
Glaucopsyche lygdamus	.17[a,d]	.91[a,d,e]	2	55	Ehrlich and Ehrlich 1978
Hypaurotis chrysalus	.29[a]	1.33[a]	3	59	Scott 1974a
Lycaena arota	.02[a]	1.02[a]	2	60	Scott 1974c
L. xanthoides	.10[a,d]	1.10[a,d]	2	26	Scott and Opler 1975
Plebejus icarioides	.11[d]	.93[d,e]	2	46	Ehrlich and Ehrlich 1978
P. saepiolus	.07[d]	.91[e]	2	37	Ehrlich and Ehrlich 1978
Hesperiidae					
Atelopedes campestris	.07	1.07	2	55	Burns 1968
Epargyreus clarus	.41	1.44	3	62	Burns 1968

Erynnis tristis	0[a]	1.00[a]	1	52	Shields 1967
Euphyes vestris	.36[d]	1.45[d]	3	46	Burns 1968
Goniurus proteus	.40	1.48	4	67	Pliske 1973
G. simplicius	.30	1.39	4	60	Pliske 1973
Hesperia pahaska	.13[a]	1.15[a]	3	77	Scott 1973a
H. sassacus	.28[d]	1.34[d]	3	31	Burns 1968
Lerema accius	.53[a]	2.03[a]	6	67	Burns 1968
Ochlodes snowi	.37[a]	1.65[a]	5	64	Scott 1973a
Poanes viator	.06	1.06	2	79	Burns 1968
Polites mystic	.24[d]	1.33[d]	4	49	Burns 1968
P. sabuleti	.03[d]	1.03[d]	2	35	Burns 1968
Pseudocopaeodes eunus	.04[d]	1.04[d]	2	26	Shields 1967
Thymelicus lineola	.09	1.09	2	54	Burns 1968
Wallengrenia otho	.11	1.13	3	174	Burns 1968

a. Frequency of remating probably underestimated, since more than 10 percent of females had no spermatophores.
b. Female attracts male with a pheromone.
c. Three spermatophores per female is common.
d. Sample size is less than 50.
e. There was evidence of absorbed spermatophores in the females' reproductive tracts.

Graham et al., in Pease 1968, nor Dustan 1964 found density effects in the moths *Pectinophora* and *Grapholitha*). The reasons for these differences are not clear, but they obviously mean that we must have large samples from a variety of habitats and seasons before making confident statements about remating frequencies; few of the data in Table 8.1 meet these criteria.

Taking into account these imprecisions and tendencies toward underestimates, the data in Table 8.1 fail to eliminate the possibility of sexual selection by female choice. On the contrary: remating is common, even in moth species in which females must invest both time and energy in attracting males from a distance with phero-mones. Remating definitely occurs in all species of Lepidoptera for which apparently adequate field data are available, except in *Prion-oxystus robiniae* and perhaps *Lymantria dispar*. Unfortunately there are no comparative studies of the genitalia of *Prionoxystus* (J. Franclemont, personal communication) to test the prediction that absence of remating would result in lack of rapid and divergent genitalic evolution. It is known that the male genitalia of this group are quite simple compared with those of many other lepidopterans, but this simplicity is probably primitive rather than derived (J. Franclemont, personal communication).

Field observations of *Lymantria dispar* show that the flightless females nearly always mate only once (Doane 1968). Male genitalia in this genus are species-specific (J. Franclemont, personal communication), so this can be seen as a contradiction of the sexual selection by female choice hypothesis. However, the females, which attract males with pheromones, often draw several males almost immediately (a typical example was four males in ten seconds, Doane 1968). In a situation of such high male density the females could favor some males over others by allowing coupling and/or intromission to occur more or less rapidly (see previous chapter). In addition, mated females did not always lay fertilized eggs (Doane 1968), suggesting that some aspects of this story are still not com-pletely understood. In sum, this case is not definitive evidence against the female choice hypothesis.

Heliconius *butterflies: a test case*

The neotropical butterfly genus *Heliconius* is particularly interest-ing because male genitalia are relatively complex and species-specific in some species groups but not in others. Emsley (1965) reviewed the taxonomy of *Heliconius* and provided clear and specific evaluations of the species specificity of male genitalic form for each species. This genus has also been intensively studied from several other points of view (see review by Brown 1981), and the frequency of mating by females is known to vary; in some species females are mated as they first emerge from their pupal skins by a waiting male, and they tend not to mate again; in others the females mate later and remate much more frequently.

The female choice hypothesis predicts a positive correlation between remating frequency and the degree of species specificity in male genitalia, and Table 8.2 shows a trend in the predicted direction. While this support is encouraging for the female choice hy-

Table 8.2 Correlation between female tendency to remate and species specificity of male genitalic form in the butterfly genus *Heliconius*. (Data on mating frequency from Brown 1981; data on genitalia from Emsley 1965.)

Most females mate only once	Most females remate	Most females mate only once	Most females remate
Male genitalia not species-specific		Male genitalia species-specific	
charitonia	*cydno*[a]	*erato*[c]	*alipherus*
sarae	*pachinus*[a]	*hermathenae*[c]	*isabellae*
leucadius	*hecale*		*melpomene*
antiochus	*elevatus*		*doris*
sapho	*ethillus*		*egerius*
hewistoni	*atthis*		*burneyi*
hecalasius			*wallacei*
telasiphe			*natteri*
clysonymus			*tales*
hortense			*lybius*
xanthocles			
vibilius[b]			
pavanus[b]			

a. The species *H. cydno* and *H. pachinus* diverged very recently (Brown 1981); taken as a unit they are distinct genitalically from other species in the same species group.

b. Male genitalia are less elaborate than those of closely related species, such as *isabellae,* that remate more often.

c. Emsley (1965) says of *erato* and *hermathenae* that "male genital valves are a good character" (pp. 214, 215), but later (p. 243) he says they are "basically similar" to those of *hecalasius* and two other species.

pothesis, there are a number of complicating factors: genitalic movements involving morphologically similar structures may vary between species; females may discriminate among males on the basis of cues other than those furnished by male genitalia (for instance, chemicals from specialized male wing scales); and even early-mating females sometimes remate (for instance, 8 percent of 120 *H. erato* females caught in the wild had remated; and some *H. erato* females lacking spermatophores were not virgins, indicating that spermatophore counts in this species give underestimates of mating frequencies: Ehrlich and Ehrlich 1978; see also Table 8.1). Thus a general rather than a strict correlation is all that can be expected. If similar tests on other groups of animals give comparable results, they will together constitute strong evidence favoring this hypothesis.

It is noteworthy that the relative simplicity of the genitalia of early mating species and early mating itself are probably derived rather than ancestral characters (Emsley 1965, Brown 1981); thus genitalic reduction has evolved concomitantly with reduced frequency of female remating in *Heliconius*.

Sex ratios and sperm exhaustion

Strong evidence regarding the frequency of female remating in species without persistent spermatophores may come from sex ratios. As Hamilton (1967) showed, there is a strong association between female-biased sex ratios and the tendency to inbreed. If females of these species frequently remated after leaving their siblings, it would be advantageous for them to produce more sons and thereby reduce or eliminate the female bias in their broods. Thus strong female bias suggests both sib mating and lack of remating after females leave their sibs. The female choice hypothesis predicts that inbreeding species with female-biased sex ratios and in which females copulate only once with their sibs should have relatively simple and uniform genitalia. This appears to be true. The largest group with this breeding pattern is the parasitoid wasps (Hamilton 1967; Askew 1968), and in general the genitalia of these wasps are simple and uniform (Askew 1968). In some eulophid genera, male genitalia are species-specific (Miller 1970), but mating away from the emergence site does occur in some groups (J. Waage, personal communication on *Trichogramma*), so these eulophids will have to be checked for remating.

In any case, the correlation between inbreeding and simple genitalia is not expected to be perfect, since there could be competition among brothers if some females mated with more than one of them. This may explain the male morphology of pymotid mites, in which inbreeding is especially clear since offspring often mate while still inside their mother (Cross 1965). Although males of many species in this group are rare and as yet uncollected, it has been found that the male structure that apparently receives the female opisthosoma during copulation presents a variety of forms and surface textures (Cross 1965). This structure may prove to vary between species.

Because sperm cells are relatively cheap, there must be strong selection on males to provide more than enough sperm in an ejaculate to fulfill the possible needs of the female during the entire lifetime of the sperm within her (see Parker 1970). In many cases the sperm can live as long as the female. Only if females normally mate repeatedly will this selection on males to "fill up" their mates be relaxed. I would thus argue that if a female's store of otherwise long-lived sperm from a single mating is exhausted before she finishes reproducing, it is probably a good indication that females of that species are normally inseminated more than once. Scattered reports of sperm reserves being commonly exhausted include the following groups (the list is undoubtedly far from complete): acarid mites (Woodring 1969; Griffiths and Boczek 1977), the lizard *Hemiergis* (Saint-Girons 1975), the poeciliid fish *Lebistes reticulatus* (Clark and Aronson 1951), pulmonate snails (Duncan 1975), the bug *Oncopeltus* (Economopoulos and Gordon 1972), *Rhagoletis* flies (Nielson and McAllan 1965, in Prokopy et al. 1971), *Drosophila* spp. (Lefevre and Jonsson 1962; Gromko and Pyle 1978), the moth *Atteva* (Taylor 1967), and the Mediterranean fruit fly *Ceratitis* (Nakagawa et al. 1971, in Walker 1980); in *Ceratitis,* females apparently lose sperm even when they do not lay eggs (Cunningham et al. 1971). Evaluation of these data is complicated, however, by the fact that females in nature do not usually live as long as those in captivity, and sperm depletion in some cases may be an artifact of captivity (Austad 1982).

Summary

It is not easy to specify the number of matings by females that would make sexual selection by female choice an important enough force to produce rapid and divergent genitalic evolution. Certainly the more males a female contacts, and the more consistently she differentiates among them with respect to the fertilization of her eggs, the stronger will be the competition between males and the stronger the possible selection on their mating structures. Females of some of the lepidopterans in Table 8.1 appear to remate rarely, perhaps less than 10 percent of the time in some species. These females must have a much smaller degree of choice among males than occurs in leks, where sexual selection by female choice is classically thought to be especially important. There are reasons to believe that the data in Table 8.1 are underestimates, but it is nevertheless reasonable to criticize the female choice hypothesis on the grounds that in such situations sexual selection would produce changes much more slowly than in leks, and that it may not produce changes quickly enough to account for the typically rapid evolution of genitalia. On the other hand, the changes in genitalia produced by sexual selection are less likely to be slowed by the action of natural selection in other contexts than are changes in other characters (see Chapter 5), and may thus be more continuous through time. Furthermore, natural selection may tend to maintain females' abilities to respond to male genitalic stimuli even when there is no genetic variance among males for genitalic characters, and males will also continue to compete, regardless of whether there is any selection on females to "choose" them (West-Eberhard 1983). Thus relatively rapid rates of genitalic evolution under sexual selection by female choice are not ruled out.

A strict prerequisite of sexual selection by female choice is that females must mate with — or have genitalic contact with — more than one male. The hypothesis could be disproved by showing strict monandry in the field for several species in a group in which genitalic differences among species are pronounced compared with differences in other characters. We have seen in this chapter that data on remating frequencies in the field are absent for most species, and incomplete for most of the others. What data there are suggest that the common idea that most females mate only once is mistaken. In short, not very much is known, but the information available does not rule out the female choice hypothesis.

9 *Apparent contradictions of the female choice hypothesis*

Several types of data, some of which were presented in previous chapters, appear to conflict with predictions of the sexual selection by female choice hypothesis. This chapter examines these data and shows that in general they are compatible with the theory. In some cases apparent exceptions are actually in accord with the predictions of the hypothesis.

Interspecific differences in female genitalia

The female choice hypothesis supposes that males use their genitalia to court females, not vice versa, so groups in which female genitalia are more diverse than those of males might seem to contradict the theory. This applies to, among others, beetles such as *Ptomophagus* (Peck 1973), *Bambara* (Dybas and Dybas 1981), and alticine chrysomelids (Samuelson 1973); some trichopterans (Nielsen 1970); proturans (Tuxen 1970); and *Scytodes* spiders (Valerio 1981). However, this pattern is not necessarily contrary to the predictions of the female choice hypothesis. Males of different species could move a particular structure in different ways to produce different stimuli in females, and females could discriminate among males via modifications of their own genitalia. Surprisingly complex movements of male genitalia have been documented, for instance in a vespid wasp (West-Eberhard 1984; see also next chapter). Another possible factor is that some species have elaborate soft or hidden male genitalic parts that are more divergent than the easily examined hard parts classically used in taxonomic studies. The plant beetle *Altica*

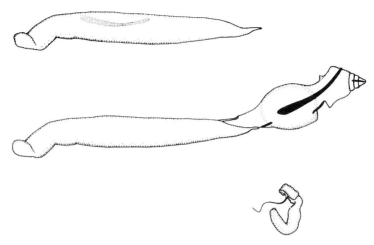

Figure 9.1 Lateral views of male genitalia at rest, *above,* and expanded, *middle,* and of the female spermatheca, *below,* from the chrysomelid beetle *Altica* sp. All are drawn to the same scale. Although the male aedeagus is quite simple when at rest, a system of membranes and six sclerites (solid black structures; the two central sclerites are paired) assumes a complex and stereotyped shape when inflated. Some of the sclerites can be seen inside the aedeagus when at rest and have been used in some taxonomic studies (White 1968). Some taxonomic studies of *Altica* have used only the external resting morphology of the male genitalia (Scherer 1969; Samuelson 1973), thus underestimating their complexity and also probably their taxonomic usefulness. The complex shapes of the female spermathecae show greater intergeneric differences than do the male genitalia (Samuelson 1973). There is no sign of ''hand-in-glove'' fit with male structures even when they are expanded.

sp. (Fig. 9.1) illustrates this possibility. The resting aedeagus is quite simple and of limited use taxonomically (Samuelson 1973). But inside it are at least six separate sclerites embedded in a membranous system which, when inflated, assumes a complex form. This may be a common phenomenon in some insect groups, since other beetles, including several chrysomelid species (White 1968, and personal observation), the silphid *Garytes* (Mroczkowski 1966), the bruchid *Acanthoscelides* (Johnson 1983), the scaphidiid *Baeocera* (Lobl 1977), and the lice *Enderleinellus* (Kim 1966) and *Austromenopon* (Price and Clay 1972) (see Fig. 3.3) all have sclerites similar to those of *Altica* inside their aedeagi; the weevil *Rhinospathe albomarginalis* also has an inflatable membranous aedeagus of complex form (Eberhard, unpublished data). Jeannel (1941) also noted that a chitinized sac could be everted from the aedeagus of the hydraenid beetle *Meropathus,* and Hubbell and Norton (1978) found that the phallus of the cave cricket *Hadenoecus opilionoides* was more complex when everted. The taxonomic usefulness of everted male organs has already been demonstrated in other groups (on polygyrid snails, Webb 1947; on snakes, Dowling and Savage 1960; Klauber 1972; and Myers 1974; on spiders, Comstock 1967; on lacewing insects, Tauber 1969; on plecopteran insects, Picker 1980; Zwick 1982; on cydnid bugs, Dolling 1981). So in some groups the apparent simplicity and uniformity of male as compared to female genitalia may be an artifact of incomplete study.

Still another possibly important factor is suggested by the work of

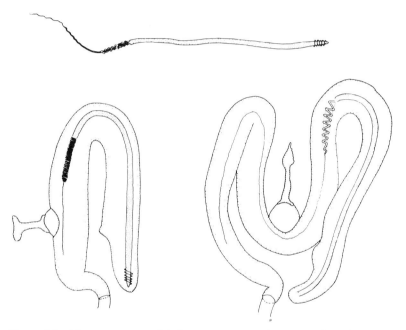

Figure 9.2 Above, sperm and, *below,* two spermathecae, each with a single sperm inside, of three different species of the featherwing beetle *Bambara.* The giant sperm (some are two-thirds the length of the beetle) vary in overall length as well as in the relative length of the head; in the length, spacing, and number of turns of the screwlike structures on the head and in the middle region; and in the length of the tail. Shapes and sizes of the spermathecae are also species-specific. The inside diameter of the spermathecal duct (apical portions shown at bottom) approximates the diameter of the sperm, and the length of both the duct and the lumen of the spermatheca correlate closely ($r =$ .98 and .94) with the approximate diameter of the sperm, suggesting a functional morphological relationship between sperm and spermatheca. (After Dybas and Dybas 1981.)

Dybas and Dybas (1981) on the featherwing beetle *Bambara.* The diameter and length of the spermatheca and the spermatheca duct furnish good species characters, and these correlate with the dimensions of the giant sperm cells of the males. Sperm size and morphology (principally the size and location of screw-threadlike structures) are also good species characters (Fig. 9.2), while the male genitalia are relatively invariable. Dybas and Dybas suggested a lock and key explanation for sperm-spermatheca correlations; this may or may not be correct — they found no sites in the spermathecae corresponding to the screw-thread structures, and they did not eliminate alternative possibilities such as stimulation from the sperm or exclusion of other sperm (Thornhill and Alcock 1983). The point is that forces of selection may sometimes be on the sperm rather than on the male genitalia and that changes in female genitalic structures could sometimes result from changes in male sperm. Dybas and Dybas (1981) cite studies showing differences in the sperm morphology of closely related species of rodents, bats, and birds.

Lack of species specificity in genitalia

The conditions for the operation of sexual selection by female choice on genitalia are nearly universal: the male genitalia must fit with or stimulate the female genitalia or must trigger female reproductive processes such as gamete transport and inhibition of remating. Therefore it seems difficult to explain why in some groups male genitalia fail to provide useful characters at the species, or even higher, level. This is the case in termites (Roonwal 1970), aphids and coccid homopterans (Ossiannilsson et al. 1970; Ray and Williams 1980), some crickets (Alexander and Otte 1967), many parasitic wasps (Askew 1968), amphipods (Bousfield 1958), and many turtles (Zug 1966), as well as innumerable small groups with scattered taxonomic affinities.

Two factors mentioned above could explain some instances of lack of genitalic differentiation. If females never make genitalic contact with more than a single male, sexual selection by female choice cannot occur, and this may explain some groups such as the parasitic wasps and termites (see previous chapter). Another consideration is the possibility of female choice based on nongenitalic cues, such as premating courtship. If the only cues used are nongenitalic male characteristics, such as coloration, courtship behavior, or substances in the semen, then the genitalia themselves would not evolve rapidly and divergently. In effect they could be sheltered from sexual selection by prior strong screening according to other criteria.

It is difficult to test the predictions because we know so little about the cues used by females and because there is no reason why they would consistently use only one cue or even one type of cue. This flexibility is illustrated in the *Argia* damselflies discussed in Chapter 2, in which colors, clasping structures, and penis structure all are species-specific in many cases (see Fig. 2.3). Even if a species has some nongenitalic character or characters that may be subject to sexual selection by female choice, this does not mean that sexual selection will necessarily fail to operate on male genitalia. The resulting prediction is thus very weak: in some, but not all, such species male genitalia will not be useful species characters. This does seem to be true, as shown by the data in Tables 2.1 and 2.2.

There are still, however, some unexplained cases. For instance, in argasid and ixodid ticks the male clearly stimulates the female to relax the opening to her vulva by rubbing his mouthparts (hypostome and/or chelicerae) inside the female genital orifice for some period of time, causing it to gradually swell before introducing a spermatophore with his mouthparts (Nuttall and Merriman 1911; Evans et al. 1961; Gladney and Drummand 1971; Feldman-Muhsam 1973; El Said 1976; Graf 1978). Yet the male mouthparts generally differ less among species than do those of females (female mouthparts are important characters for distinguishing species) (B. OConnor, personal communication). Some female ticks remate several times (Nuttall and Warburton 1911), so sexual selection by female choice would be expected to occur. It is possible that males stimulate females chemically in these species; males of some species have specialized salivary glands, and salivation occurs while the male mouthparts are inside the female's vagina (El Said 1976), but these secretions may have only a simple lubrication function (Feldman-Muhsam et al. 1979).

It is noteworthy that the species specificity of the female mouthparts in ticks fits a pattern found in organs other organisms use to anchor themselves to hosts and prey: in tapeworms the scolex is often species-specific (Joyeux and Baer 1961; see Fig. 9.3); coelenterate nematocysts are very diverse and complex (Hyman 1940); the mouths of many nematodes are useful in distinguishing species

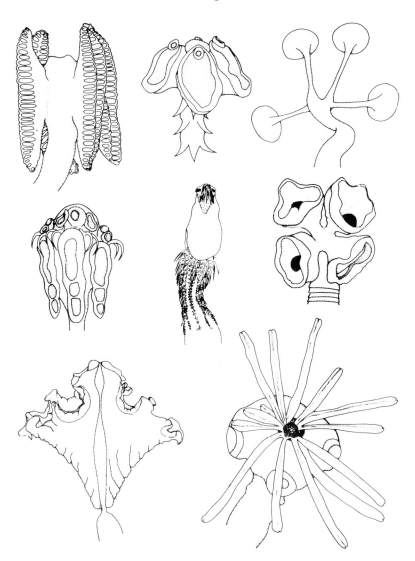

Figure 9.3 A sampler of scolex (holdfast organ) morphology in tapeworms, like an array of sinister flowers (drawn to different scales). As with nematodes, trematodes, acanthocephalan worms, pentastomids, and ticks, the holdfasts of tapeworms are often complex and species-specific in form. Possibly hosts and parasites are engaged in evolutionary races with respect to the parasite's ability to attach to the host. This might explain the lack of secondary sexual differentiation of male tick mouthparts, despite their use as stimulators in courtship. (*Upper left*, after Carvajal 1977; *center*, after Williams and Campbell 1980; others after Joyeux and Baer 1961.)

(Hyman 1951a); the holdfast opisthaptors of some monogenic trematodes are complex and species-specific in form (for example, Paperna 1972; Kritsky and Thatcher 1974); the holdfast organ, or claw, of pentastomids is often species-specific (H. W. Levi, personal communication); and the spiny head of acanthocephalan worms assumes a wide variety of forms and is a useful taxonomic character (Hyman 1951b). Perhaps these parasites and their hosts are engaged in evolutionary races, and each holdfast design is best for each parasite's host or collection of hosts; in that case differentiation in the structure of male ticks' hypostomes would be disadvantageous. This explanation is not very convincing for ticks, however, since most genera deposit an amorphous sheath of "cement" in the wound they make in the host, and then embed the hypostome in this cement (Kemp et al. 1982); the functional significance of species differences in mouthparts is unclear.

High paternal investment in offspring

The correlation between sex role reversal and the presence of intromittent organs in females rather than males was discussed in Chapter 5 to justify the female choice hypothesis. There are a number of exceptions to this correlation. In belostomatid bugs males rather than females care for the offspring, but males have intromittent genitalia that are species-specific in form (Smith 1979; Menke 1960), and male *Apogon* fish introduce sperm into females and care for the offspring (Blumer 1979). Descriptions of the mating behavior of these species make it clear, however, that despite — and in fact because of — the large paternal investment in their offspring, the males are probably under stronger selection than the females to make certain that copulation and fertilization occur. In contrast to the seahorses and pipefish, fertilization occurs inside the female in both *Apogon* fish and belostomatid bugs, and the males thus run the risk of caring for offspring sired by previous mates of the females. In both groups the males repeatedly solicit and mate with the females, apparently to swamp any sperm remaining from previous matings (Blumer 1979; Smith 1979). Only if it were shown that the intromittent genitalia of the males of these groups evolved *after* paternal care arose (a possibility in *Apogon,* but not in the bugs) would the hypothesis be clearly contradicted. But as the data stand, there is no conflict.

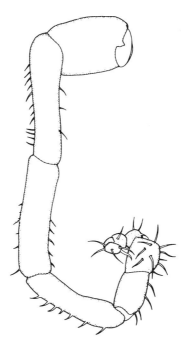

Figure 9.4 The oviger (modified leg) used by male pycnogonids to gather up and carry eggs. Oviger morphology is especially useful for distinguishing closely related species of these distant relatives of spiders; why these organs should follow the typical genitalic pattern of rapid divergence is not clear. (After Hedgpeth 1963.)

One group with male parental care and species-specific male secondary sexual characters remains enigmatic. These are the pycnogonids, a little-known taxon of marine chelicerate arthropods. There are essentially no external genitalia in either sex, and fertilization is external. The male presses his gonopore close to that of the female, and fertilizes the eggs as they emerge (King 1973). The male then immediately gathers up the eggs with his ovigers (modified legs present only in mature males — see Fig. 9.4) and carries them for a variable period of time (King 1973). Ridley (1978) suggested that the ovigers touch the female during the oviposition process, but this seems unlikely, judging from the descriptions I have seen in original reports (Cole 1901; King 1973). Pycnogonid ovigers have evolved in the typical pattern for genitalia and are such critical characters for species identification that some species cannot be identified otherwise (Hedgpeth 1963). If males do indeed touch females with their ovigers during oviposition, it is possible that they serve as stimulatory organs, and their evolution could be explained by sexual selection by female choice. Otherwise none of the hypotheses seem applicable. More detailed observations of the reproductive biology of these organisms would be of great interest.

Intraspecific uniformity of genitalia

The female choice hypothesis predicts that genitalic structures in isolated populations or subspecies are likely to diverge, and thus a certain amount of intraspecific variability in genitalia is expected. This seems to conflict with the fact that taxonomists use genitalic structures to typify species, implying that these structures are uniform within many species.

This argument is difficult to evaluate. Sexually selected characters do often show particularly great intraspecific variations (Darwin 1871; West-Eberhard 1983), but many, such as plumage colors of some birds, dewlaps in *Anolis* lizards, and the calls of frogs and crickets, are nevertheless especially useful in species-level taxonomy (Mayr 1963; West-Eberhard 1983). To be a useful taxonomic character, a structure need only be *relatively* invariable intraspecifically — that is, vary less within the species than it usually varies between species. The widespread use of genitalia in taxonomy is thus an indication of relative rather than absolute invariability.

In fact, the clear lack of genitalic uniformity has been demonstrated in studies like those of Levi (1968, 1971, 1974, 1977, 1981) on araneid spiders, in which the size and shape of male and female genitalia clearly vary intraspecifically; (see also Fig. 9.5). In this group most taxonomic papers give "the" form of the genitalia for each species and, as has been traditional for araneid taxonomists (Archer 1951a, b; Gertsch 1964; Grasshoff 1964, 1968) Levi (1978) puts great emphasis on genitalic characters in making taxonomic decisions. Intraspecific variation in genitalia that are nevertheless species-specific in form is known in other groups besides these spiders: other spiders (Coyle 1969, 1971, 1974, 1981; Hippa and Oksala 1983), homopterans (Fennah 1946; Kunze 1959; Taylor 1962), hemipterans (Kerkis 1931), coleopterans (Kistner 1966; Peck 1984), lepidopterans (Jordan, in Goldschmidt 1940; Lindsey 1939; Lorkovic 1952; Turner et al. 1961; Shapiro 1978), a dragonfly (Johnson

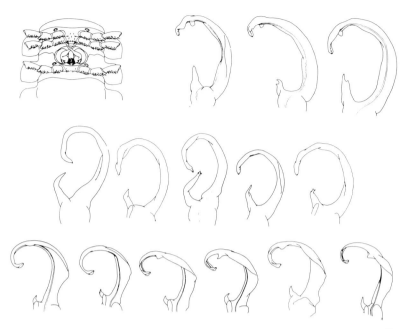

Figure 9.5 Intraspecific and intrasubspecific variation in the male genitalia (gonopodial telopodites) of *Sigmoria* millipeds. *Top row, S. nigrimontis intermedia; middle row, S. n. nigrimontis; bottom row, S. stenogon.* Gonopod characters are species-specific throughout the genus but are also intraspecifically variable. (From Shelley 1981.)

1972), millipeds (Shear 1972; Shelley 1981; — see Fig. 9.5), mites (Pritchard and Baker 1955), crabs (Crane 1975), rattlesnakes (Klauber 1972), nematodes (Gibbons 1978), primates (Hershkovitz 1979), bats (Pine et al., 1971), and poeciliid fish (Gordon and Rosen 1951). Taxonomists, being generally more interested in typifying different species to permit their identification than in documenting the species' ranges of variability, probably have tended to focus on and thus overemphasize the uniformity of genitalia within species. The common belief that genitalia function as species recognition devices may have also prejudiced perceptions of genitalic variability.

In addition, taxonomists may have sometimes put too much emphasis on genitalia as species characters. This could lead to too much splitting of species, with the genitalic characters being relatively invariable within the subspecific groups that the taxonomist recognized as species. Intraspecific invariability in the genitalia of some groups could thus be an artifact of overreliance on genitalic characters.

In sum, the claim of intraspecific invariability in genitalia is probably too strong. It is more nearly correct to say that genitalia are generally neither invariant nor so variable that the ranges of variation of closely related species come near to overlapping. This intermediate degree of uniformity is not clearly in conflict with the sexual selection model.

Spermatophore complexity and male-female contact

The tendency for male-female contact to be associated with complex spermatophore morphology (Table 3.4) was cited in Chapter 3 as evidence against the species isolation hypotheses. It could also be interpreted as evidence against the sexual selection hypothesis: if females discriminate among males, it would seem especially important for males that never encounter their prospective mates to leave stimulating, "persuasive" spermatophores. This objection fails to take into account that under the female choice hypothesis, complex spermatophores are taken to be stimulators of, or to mechanically mesh with, female *genitalia*. Males that do not contact females cannot oblige them to be inseminated — a female of such a species does not touch her genitalia to the spermatophore unless she is receptive. On the other hand, in a species in which males contact females and court them before depositing spermatophores, males will sometimes maneuver less receptive females into positions in which spermatophores are in the near vicinity of the female's genitalia. Both attended and unattended spermatophores can be sensed by marginally receptive females, but only attended spermatophores are likely to stimulate and/or fit into such a female's genital area. Any characteristic of a spermatophore, such as a guiding spine or an injecting mechanism, that makes it more likely that the female's genitalia will actually take up the sperm, would be favored. Female preference for such characters could then be additionally favored because of the superior sons that would result. That such "persuasive" structures on a spermatophore can be important is illustrated dramatically by Arnold's (1972) observations of four species of *Amblystoma* and *Plethodon* salamanders. The males lure females

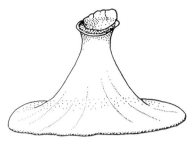

Figure 9.6 The "inert" spermatophore of the salamander *Ambystoma macrodactylum*. The naked sperm mass sits at the tip of a pedestal; after the male courts the female, she seizes the sperm mass with her cloaca. But sometimes a female is not completely receptive and lowers herself onto a spermatophore without taking up the sperm. This illustrates the advantage of "active" spermatophores that inject sperm into the female (as in scorpions, pseudoscorpions, and squids). (After Anderson 1961.)

toward spermatophores (Fig. 9.6) they have deposited, and sometimes a female lowers her cloaca onto a spermatophore but fails to pinch off the packet of sperm at the tip, leaving the sperm mass apparently intact when she moves on (Shoop [1960] probably saw the same phenomenon in *A. talpoideum*). In addition, in those species without male-female contact, the apparent competition among males to lay greater numbers of spermatophores could favor reduction in spermatophore complexity to enable the male to produce greater numbers. When one takes into account the differences in receptivity of females whose genitalia come into contact with spermatophores plus the possibility of male-male competition, the general trend in Table 3.4 is in agreement with the female choice hypothesis.

There are several exceptions to the pattern in Table 3.4, for example the collembolans and pionid mites with male-female contact but simple spermatophores and other mites without contact but with complex spermatophores. In the collembolans, at least, sexual selection by female choice may have focused on the male clasper organ (his antenna) rather than the spermatophore (see Table 11.1), but in the absence of more complete data on points such as females' abilities to distinguish spermatophores of other species from those of their own, and the phylogenies of the groups involved (and thus the derived versus primitive nature of the spermatophore morphologies and the male-female behaviors), speculation is fruitless.

Infrequency of postcopulatory courtship

This difficulty is more theoretical. The sexual selection by female choice hypothesis focuses on the female's ability to determine the fate of sperm through postcopulatory reproductive processes. If such processes are critical, however, we would expect males commonly to court females after insemination until one or more of the processes (for example sperm transport) was carried out. The apparent infrequency of postcopulatory courtship in animals seems not to be in accord with this prediction.

It is difficult to evaluate this objection. Postcopulatory courtship does occur in some species: stroking in the bee *Centris pallida* (Alcock and Buchmann, in press), vocalization in rats and mice (Sachs and Barfield 1976), offering of chemical secretions in some millipeds (Haacker 1974) and in a tree cricket (Gywnne 1983), tapping with antennae, head, and abdomen in the wasps *Parancistrocerus pensylvanicus* (Cowan, in prep.) and *Abispa ehippium* (Smith and Alcock, in Cowan, in prep.), and antennation in the beetle *Cryptolestes pusillus* (Wocjik 1969). The postcoital display of male uganda kob is thought to stimulate female transport of gametes (Halliday 1980). In the earwig *Metresura flavipes* male courtship continues many minutes after copulation has begun (personal observation). Few studies of copulatory behavior have established the precise moment when insemination occurs; the later stages of some copulations may occur after insemination and thus represent postinsemination courtship. Thus the phenomenon does occur (and does not have any other obvious explanation in most cases), and it may be more widespread than is thought. In our present state of ignorance it may not be possible to judge even in qualitative terms, how common postcopulatory courtship is, so it does not provide a compelling reason to reject the female choice hypothesis.

Patterns of genitalic change

The female choice hypothesis postulates accumulations of small changes over time rather than abrupt changes followed by long periods of relative stasis, but some data from fossil insects show long periods of stasis that appear to contradict this gradualist scheme. Coope (1979) showed that evolutionary lines of recent fossil insects whose genitalia are sufficiently intact to permit species identification often show very little genitalic change over relatively long periods of time. The groups (all from the temperate zone) for which this type of evidence is available have gone through repeated, abrupt, and radical changes in geographic ranges in the past several million years, and Coope argued that the range changes are responsible for the lack of changes in the species. Such range alterations presumably mean that isolated populations either go extinct or merge with other populations before they have differentiated enough to become reproductively isolated. This explanation is compatible with the female choice hypothesis as outlined here (for a contrasting view of the lack of change in fossil species, see Olson 1981).

Summary

This chapter represents a minefield for the female choice hypothesis. Some of the apparent problems, such as species with high male parental investment and species-specific male genitalia, and the correlation between spermatophore complexity and direct male courtship of females, turn out to not really be problems. In fact the hypothesis actually predicts these correlations, once critical points are clarified. Other patterns — the fact that interspecific variability in some groups is greater in female than in male genitalia, the lack of genitalic differences between some species, excessive genitalic uniformity within some species, the general lack of postcopulatory courtship, and stasis in fossils — can be explained by the hypothesis, but the data necessary to test these explanations are lacking. Two groups (ticks and pycnogonids) remain mysteries to me. It is worth noting that nearly all of the remaining problem topics involve little-studied groups and phenomena; perhaps future discoveries will clarify presently obscure points.

10 *Use of genitalia as stimulators*

Two major predictions of the female choice hypothesis relate to the ways genitalia are used during copulation and to the pattern of evolution of nongenitalic male organs used to contact females. This chapter and the next show that both of these predictions are in accord with available data.

According to the female choice hypothesis, genitalia are devices that are often used to stimulate the female. It predicts that both genitalic form and behavior should often give signs of being designed to accomplish this function. It also predicts that those parts of the genitalia that stimulate the female should evolve rapidly and divergently and thus be species-specific in form. We are almost totally ignorant of exactly how male genitalia move and mesh within female genitalia in most cases, but several kinds of data do lend support to these predictions.

External movements

In the carabid beetle *Pasimachus punctulatus* the male taps his parameres rhythmically against the surface of the female's abdomen during copulation (Alexander 1959). The parameres are especially useful in distinguishing species of this genus (Banninger 1950), as they vary in length and breadth as well as density of hairs. In the lepidopteran genus *Erebia* the male genitalic valves (Fig. 10.1) continually rub against the female's abdomen during copulation (Lorkovic 1952). The spines and teeth on these structures, which are species-specific in form, are thus used not to hold the female but

Figure 10.1 The male genitalia of a satyrid butterfly. The "valves" *(c)* are often species-specific in form and were thought to function as claspers. But Lorkovic's careful observations of copulating pairs showed that they are stimulatory organs; the male rubs them constantly back and forth on the surface of the female's abdomen during copulation. (After Lorkovic 1952.)

instead to stimulate her. Scott (1978) observed a similar use of male valves in the hesperiid butterfly *Erynnis persius,* and again the male valves in this group are useful for distinguishing species (Burns 1970). Platt (1978) observed that in the nymphalid butterfly *Limenitis* the male distal teeth and hooks are raked across the lateral abdominal sterna of the female during copulation, and again these structures are good species characters (Platt et al. 1970). In the cocinellid beetle *Cycloneda sanguinea* the male's parameres remain outside the female during copulation and he repeatedly taps them against her abdomen (O. Trejos and W. Eberhard, unpublished); these structures (Fig. 10.2) have species-specific forms in this genus (R. Gordon, personal communication), and also remain outside the female during copulation in other species (R. Gordon, personal communication). Although male clasper morphology does not suggest a stimulatory function in damselflies (Fig. 5.4), females in the genus *Enallagma* have species-specific patterns of receptors that are stimulated when the female is clasped by the species-specific male clasping organs (Robertson and Paterson 1982). Experiments have shown that female *Lestes* and *Ischnura* damselflies apparently sense the species identity of males via the clasping organs (Loibl 1958; Krieger and Krieger-Loibl 1958). Male bursae (grasping organs) of nematodes are often used to distinguish species, and the bursa of the male *Nematospiroides dubius* is moved during copulation (Croll and Wright 1976).

It is also not uncommon for male insects to tap females near the genital orifice prior to intromission. An undoubtedly incomplete list of insect species with species-specific genitalia that do this includes the beetles *Pasimachus punctulatus* (Alexander 1959), *Tribolium confusum* (Wocjik 1969), and the wasps *Ancistrocerus antilope* and *Parancistrocerus pensylvanicus* (Cowan in prep.; references for species specificity of male genitalia are, respectively, Banninger 1950; Halstead 1969; and D. Cowan, personal communication). In some of these species the female must respond appropriately, as by opening the tip of her abdomen in some way, for copulation to occur, so the genitalic tapping apparently constitutes part of the male's courtship.

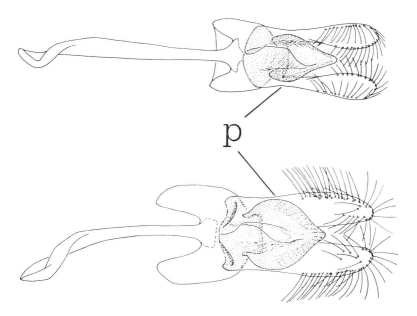

Figure 10.2 Male genitalia of two congeneric species of azyine ladybird beetles. In ladybirds the parameres *(p)*, which vary in shape in different species, have been observed to remain outside the female and tap rhythmically on her abdomen during copulation. (From Gordon 1980.)

Intromission without insemination

In several species copulation normally involves one or more insertions of the male's genitalia that do not result in insemination. In six
species in two genera of cicindelline tiger beetles, the male typically
inserts his genitalia deeply into the female, withdraws them, and
then inserts them again and ejaculates (Freitag et al. 1980). In one
species it has been shown that insemination occurs only during the
second intromission. The shape of the genitalic flagellum, which
varies between species, closely matches the shape of the female
spermatheca duct in all of these species. Since it is anatomically
impossible for the sperm to pass through the flagellum (it is not
hollow and has no hole at the tip), Freitag and colleagues proposed
that in the first insertion the flagellum was inserted into the spermatheca duct and thus "opened and prepared the lumen for sperm
passage . . . to the spermatheca." Brinck (1956a) postulated an essentially identical use for the flagellum (probably not a homologous
structure) of some plecopterans. Similar cases, in which the male of
some spiders and millipeds (Fig. 10.3) inserts his species-specific

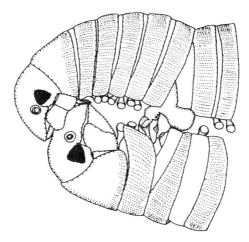

Figure 10.3 A copulating pair of the milliped *Cylindroilus punctatus* (legs
and antennae omitted). The male must charge his secondary genitalia
(modified legs or gonopodia) with sperm before inseminating the female.
But first the male inserts his gonopods into the female and moves them
rhythmically. After doing this for several minutes he withdraws them,
loads them with sperm, reinserts them, and inseminates the female. Such
preliminary intromissions may stimulate females to receive and utilize
sperm. (After Haacker and Fachs 1970.)

secondary genitalia into the female before charging them with sperm, then withdraws them and primes the genitalia with sperm before inserting them again and ejaculating, were cited in Chapter 5. Males of the moth *Lasiocampa quercus* also insert their genitalia into the female twice, first in the vagina, where sperm is deposited, and then (through a different opening — see Fig. 6.2) into the *oviduct;* the second insertion apparently serves to induce oviposition rather than to introduce sperm (Pictet, in Parker 1970). Male genitalia are species-specific in *Lasiocampa* (J. Franclemont, personal communication).

The eumenid wasps *Paraleptomenes miniatus, Abispa ephippium,* and *Ancistrocerus antilope* all copulate more than once during a single pairing (reviewed by D. Cowan, in prep.), as do the walkingsticks, *Diapheromera* (Sivinski 1978), and the beetle *Hippomelas planicosta* (Alcock 1976). It is not known when sperm are transferred; in *A. antilope,* at least, genitalia are species-specific (D. Cowan, personal communication). Males of the platyrhacid milliped *Nyssodesmus python* also perform a series of intromissions of progressively shorter duration during their one- to two-hour couplings (personal observation), and male genitalia in this genus are species-specific in form (W. Shear, personal communication). Most species of spiders probably insert their genitalia, which are usually species-specific, into the female repeatedly rather than only once (Gering 1953; Bristowe 1958). If bulb expansion in the pedipalp is a sign of sperm transfer, then in many species some insertions do not result in insemination (Robinson 1982).

Male rhodacarid mites have modified, species-specific chelicerae that they use to insert spermatophores into the female's vagina or, in some species, into other accessory "spermathecal pores" on her legs and body (See Fig. 6.7). Lee (1974) observed the mating behavior of several species and found that even in species in which the spermatophore was inserted into a spermathecal pore, the males first moved their mouthparts in the female's *vagina* as if feeding there, sometimes for more than thirty minutes. This appears to be a case in which insemination and stimulation have become at least partially uncoupled morphologically, with the insemination site changing more rapidly than the stimulation site. In *Euepicrius filamentosus* (Fig. 10.4) the male also poked the spermadactyls of his chelicerae into the female's spermathecal pores.

Males of many mammal species perform a series of more or less brief intromissions before finally ejaculating. Female rats that have

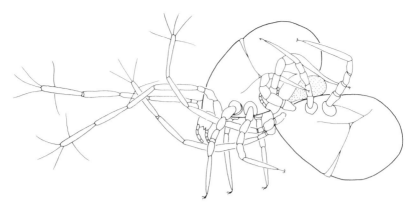

Figure 10.4 A mating pair of *Euepicrius filamentosus* mites, with the male clinging below the female as he uses his specialized, species-specific chelicerae (not visible) to introduce a spermatophore (stippled) into the spermathecal pore in the coxa at the base of one of her legs. Before depositing the spermatophore, the male appears to "feed" at the female's vagina, moving his chelicerae in and out alternately for several minutes. Then the cheliceral tips are poked alternately into the spermathecal pores on the two coxae. Only after these apparently stimulatory preliminaries does the male produce the spermatophore and inseminate the female in her leg. These mites are undoubtedly descended from ancestors in which males deposited sperm in the female's vagina, and the persistence of vaginal stimulation even when insemination is via the coxa as in this species represents a particularly clear example of the apparent importance of stimulation per se during insemination. (After Lee 1974.)

received less than four intromissions before the male ejaculates transport less sperm to the uterus and are less likely to initiate the luteal cycle, which leads to implantation in the uterus, than those which receive more than four, as is typical for the species (McGill 1977); in this case both the stimulatory use of the male genitalia and the discriminatory female response are documented.

Movements within the female

Mammalian copulations typically involve repeated male pelvic thrusting movements that cause the penis to rub back and forth inside the female. The more or less elaborate spines and denticles on the penis of many groups (cats, primates, hyenas, shrews, some rodents: Eckstein and Zuckerman 1956; Hooper and Musser 1964;

Prasad 1974; see Fig. 10.5) undoubtedly cause tactile stimulation in the female as they rub inside her genitalia. The taillike filament at the tip of a bull's penis that is flipped forward at the moment of ejaculation and withdrawal (Walton 1960, see Fig. 1.3) seems designed to do nothing other than stimulate the female. The length, position, and shape of this structure all vary in different genera of hoofed mammals (Gerhardt, in Walton 1960; Fig. 1.3). The glans of the human penis becomes more protruded during ejaculation, and several muscles contract, causing a variety of movements (Masters and Johnson 1966); a possible function for these otherwise unexplained changes is that they stimulate the female. Human penis morphology is quite different from that of other anthropoid apes (Short 1979).

Among arthropods, males of the mosquito *Culiseta inorata* *(Culiseta = Theoboldia)* and three species of spiders in the genus *Mynoglenes,* all of which have species-specific genitalia (Matheson 1929; Blest and Pomeroy 1978) move their genitalia in and out of the

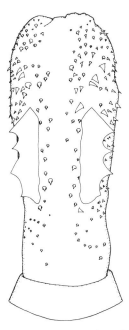

Figure 10.5 The spiny penis of a cavimorph rodent. It is difficult to imagine that the movements of such an organ go unperceived by the female while it is within her. (After Hooper 1962.)

female during copulation (Clements 1963; Blest and Pomeroy 1978) as do some acaroid mites (Griffiths and Bocjek 1977). The male of the pseudoscorpion *Dactylochelifer* shakes the female as he uses the species-specific first tarsal claw to introduce his spermatophore into her genital atrium (Weygoldt 1969). The spermatophore of the moth *Heliothis* may make corkscrew-like coiling movements of its own accord inside the female (Proshold et al. 1975). Probably many male insects move their genitalia to some extent during copulation; the beetles *Lytta, Pyrota,* and *Chauliognathus* all make repeated pumping movements with their abdomen as they copulate (Adams and Selander 1979; personal observation), and the chrysomelid beetle *Altica sp.* and the weevil *Rhinospathe albomarginalis* move their aedeagi back and forth within the female during copulation (personal observation). In *Altica* and *R. albomarginalis* the male genitalia include complex arrays of membranous and cuticular structures that are inflated during copulation (see Fig. 9.1); the complexity of these structures suggests that they may well be species-specific in form.

Still another kind of evidence comes from observations of genitalic movements in solitary, partially anesthetized males. West-Eberhard (1984) found that several parts of the genitalia of a lightly anesthetized male *Parachartergus apicalis* wasp could execute complex movements (male genitalia in this genus are species-specific, Richards 1978), and some parts that look rigid in dead specimens proved to be flexible and highly mobile. West-Eberhard noted that the genitalia made "the most fluid and subtly modulated movements I have ever observed in wasps" and concluded that "the vision of wasp genitalia as rigid insertion devices is clearly inadequate." It would be interesting to attempt to make similar observations on other organisms; findings like those of Duncan (1975), Sachs and Barfield 1976, and Masters and Johnson (1966), who noted rhythmic contractions and flipping movements of the penes of molluscs, mice, and humans, suggest that such genitalic movement may be widespread.

Indirect evidence

There are a number of additional, less direct indications that male genitalia have stimulatory functions. Perhaps the most dramatic are the devices which appear to be stridulatory organs in the male

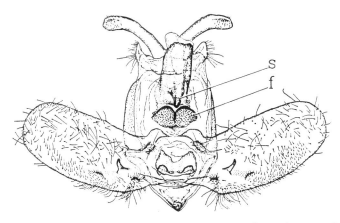

Figure 10.6 Genitalia of the moth *Olceclostera seraphica*, showing the file *(f)* and scraper *(s)* structures used by arthropods to produce vibratory signals. Female moths of this species apparently lack ears, so the males' vibratory signals probably reach her directly via her genitalia. These genitalia clearly seem to be designed to stimulate (From Franclemont 1973.)

genitalia of some apatelodid moths (Fig. 10.6, Forbes 1941; Franclemont 1973). No hearing organs have been described in this family (J. Franclemont, personal communication), so the vibrations are probably conveyed to the female directly via her genital region. The male genitalia of the vespid wasp *Chartergellus* have granulate areas that move against each other (Richards 1978) and in so doing may also produce vibrations.

In a number of groups some species-specific parts of the male stay outside the female and touch relatively featureless and interspecifically uniform parts of her body (beetles, Jeannel 1941; bumblebees, Richards 1927a); judging by their bristly or hairy morphology, the male structures may act as stimulators. Indeed, it is difficult to imagine any other function for them, and even if they had another function they would probably also act as stimulators. Many turbellarians have pairs or even rows of penes, even though there is only a single female copulatory pore; Hyman (1951) noted that the extra penes are thought to act as stimulators in copulation. The morphology of some male hymenopteran and dipteran genitalia, when combined with a basic knowledge of how male and female parts fit together, has suggested to Smith (1970) that they function as "titillators." The penes of many microlepidopterans have long, sharp, and stiff deciduous spines that the male leaves in the female bursa

(Busck 1931) and which "are of great taxonomic value" (Tuxen 1970). The species-specific arm tips of some cephalopods break off inside the female, where they remain "very active" (Wells and Wells 1977), and the embolus tips of some male spiders are found in the females' spermathecae (Abalos and Baez, 1966; Levi 1975 — see Fig. 8.2). Other male structures, often species-specific, which are called "titillators" but for which the justification, if any, of the term is unknown to me, are the paraprocts of plecopterans (Brinck 1956b), trichopterans (Nielson 1970), and some orthopterans (Ander 1970), and the ventral parameres of some mecopterans (S. Issiki: personal communication by G. Byers).

Stimulatory structures (darts and sarcobella) on the male genitalia of some gasteropod and opisthobranch molluscs are well known (Baker 1945; Fretter and Graham 1964; Beeman 1977), and species-specific penial spines are also known in some opisthobranch molluscs (Fretter and Graham 1964; Edmunds 1970). The penial spines of oligochaete worms, which are often species-specific in form, are thrust into the female's skin during copulation (Stephenson 1930; Avel 1959); they are sometimes thought of as holdfast devices but, as noted by Stephenson (1930), they may also be stimulatory devices, and at least some appear to be designed to carry gland products from the male to the female.

Finally there is the extraordinary report of photoreceptors on both male and female genitalia of several different species of butterflies (Arikawa et al. 1980). The evidence for photoreception is strong and includes gross morphology (receptors are covered by patches of hairless and transparent cuticle in the midst of hairy, pigmented areas), subcellular structure (lamellated bodies in the receptor cytoplasm), and physiology (electrical recordings from the nerves leading from the receptors show increased activity when the organs are illuminated). It was not determined how the receptors of the two sexes lie with respect to each other during copulation. I do not know of any light-emitting organs in lepidopteran genitalia. For the moment the significance of genitalic photoreceptors is an intriguing mystery.

This listing of apparently stimulatory genitalia with species-specific forms is undoubtedly incomplete. Some cases are so clear that it seems extremely likely that the genitalia are designed and used to stimulate the female in species-specific ways.

11 *Specialized nongenitalic male structures*

According to the sexual selection by female choice hypothesis, genitalia, rather than other male structures, evolve rapidly and divergently because they consistently come into contact with females in sexual contexts. Unlike other structures, they both stimulate and fit mechanically with females. In a number of animal groups, however, certain structures, such as the head, legs, or antennae of males, are specialized to contact females — often but not always to hold them — prior to or during copulation. If the sexual selection hypothesis is correct, then a clear prediction emerges: in many cases these other male structures should evolve rapidly and divergently, as genitalia do. They should often be diverse in form and more elaborate than would seem likely on the basis of their supposed function. This prediction is clearly confirmed, in many cases spectacularly.

Bizarre and species-specific nongenitalic male structures specialized to contact females are found in a wide variety of animal groups; see the undoubtedly incomplete list in Table 11.1, and Figs. 11.1 to 11.7. In fact the only case I know of in which such a modified male structure is not species-specific (there may of course be others), is in pymotid mites; one pair of legs in the male is modified to form a forcepslike structure that probably holds the female (Cross 1965). No direct behavioral observations have been made, and this case is somewhat uncertain because the males of these mites are poorly known. These mites are unusual in that they generally mate exclusively with their siblings within or near the body of their mother, so sexual selection may be less likely or may have focused on other structures, as discussed in Chapter 8).

Table 11.1 Nongenitalic male claspers and contact organs that are useful characters in species-level taxonomy.

Group	Structure	Function	Corresponding female structure also good species character?	References Function	References Taxonomy
Araneus (spider)	Tibia, leg II	Holds female femurs I and II	No	Grasshoff 1964	Grasshoff 1968
Argyrodes (spider)	Cephalothorax	Contacts female mouthparts (glandular)	No[a]	Lopez and Emerit 1979	Exline and Levi 1962
Some erigonine spiders	Cephalothorax	Contacts female mouthparts	No[b]	Bristowe 1958; Meijer 1977	Milledge 1980, 1981a,b
Euagrus (spider)	Leg II	Pins female on her back by holding leg II	No[b]	Coyle in press	Coyle in press
Some Tetragnatha (spider)	Chelicerae	Holds female chelicerae	Yes	Kaston 1948	Kaston 1948; Chickering 1959
Schizomids (arachnids)	Flagellum of telson	Female holds it with her chelicerae	No[a]	Schaller 1971	Rowland and Reddell 1979
Trouessartia (feather mite)	Hysterosoma	Touches female	No[a]	Popp 1967	Santana 1976
Glycyphagus (mite)	Comb on setae on tibiae I & II	Probably contacts female as male holds special female setae	Sometimes	B. OConnor, pers. comm.	B. OConnor, pers. comm.
Many feather mites (astigmatids)	Legs III & IV	Hold female	?	B. OConnor, pers. comm.	B. OConnor, pers. comm.
Rhodacaridae (mites)	Leg II	Probably holds female	?	B. OConnor, pers. comm.	Lee 1970
Rhinoseius (mite)	Dorsal setae	Contact female	No[a]	B. OConnor, pers. comm.	B. OConnor, pers. comm.

Organism	Male structure	Action	Female control?	Reference	Reference
Some parasotid mites	Leg II	Holds female leg IV	No[a]	Hartenstein 1962	Hartenstein 1962
Copepods	1st, sometimes 2d antenna, sometimes leg V	Hold female	?	Pennak 1978	Pennak 1978
Jaera (isopod)	Legs	Brush, press on female	No[b]	Solignac 1981	Solignac 1981
Isopods	Gnathopod	Hold female (?)	?, no[a]	Pennak 1978	Pennak 1978
Anostracan eubranchiopods	2d antenna	Holds female	?	Pennak 1978	Pennak 1978
Decapods	Periopod	Hold female	?	Pennak 1978	Pennak 1978
Chordeuma (milliped)	Dorsum seg. 16	Female licks glandular products	?	Haacker 1971	Haacker 1971
Julus (milliped)	Coxa, leg I	Female licks glandular products	?	Haacker 1974	W. Shear, pers. comm.
Symphylan collembola	Antenna	Clasps female antenna	No[a]	Mayer 1957	Massoud and Betsch 1972
Crabro (wasp)	Tibia I	Semitransparent plate held over female eyes	No[a]	Matthews et al. 1979	Bohart and Menke 1976
Mellitobia (wasp)	Antenna	Manipulates female antenna	No[a]	Evans and Matthews 1976	Evans and Matthews 1976; R. Matthews, pers. comm.
Zethus, Polistes (wasps)	Tip of antenna	Grips/rubs female antenna	No[a]	Evans and Eberhard 1970	Bohart and Stange 1965; van der Vecht 1971
Belanogaster (wasp)	Tip of antenna	Grips/rubs female antenna	No[a]	Piccioli and Pardi 1970	Richards 1982
Leioproctus (bee)	Mandible	Holds female	Yes	Toro and de la Hoz 1976	Toro and de la Hoz 1976
Some megachilid bees	Abdominal sternae and/or legs	Hold/contact female	?	Severinghaus et al. 1981; Batra 1978	Griswold 1983, G. Eickwort, pers. comm.

Table 11.1 (continued)

Group	Structure	Function	Corresponding female structure also good species character?	References Function	References Taxonomy
Apis (honeybee)	Tarsus leg III, esp. adhesive hairs	Holds female abdomen or leg III	?	Ruttner 1975	Ruttner 1975
Pasimachus (beetle)	Mandibles	Seizes female	No[a]	Alexander 1959	Bänninger 1950
Meloe (beetle)	Antenna	Holds female antenna	No[a]	Pinto and Selander 1970	Pinto and Selander 1970
Pyrota (beetle)	Antenna	Holds female antenna	No[a]	Selander 1964	Denier 1934
	Maxillary palp	Touches female elytra	No[a]		
	Leg I	Holds female leg III	No[a]		
Some Epicauta[c] (beetle)	Antenna	Wraps around female antenna	No[a,b]	Selander and Mathieu 1969	Selander and Mathieu 1969
Collops (beetle)	Antenna	Female grips with mandibles	No[b]	P. Smith, pers. comm.	Fall 1912
Boreus (scorpionfly)	Wings	Hold female, then rake across her antennae and rostrum	No[a]	Cooper 1974	Cooper 1972
Some mecopterans	Notal organ on abdomen	Grips female	Generally no	Thornhill 1980	G. Byers, pers. comm.
Ectobius (roach)	Tergal glands	Female feeds on products	?	Roth 1970	Richards 1927b
Labia (earwig)	Forceps	Grip female forceps	No	Briceño and Eberhard, unpub.	Brindle 1971
Many odonates	Abdominal appendages	Grip female head/thorax	Often	Corbet 1962	Kormandy 1959; Paulson 1974; L. Gloyd, pers. comm.

Some plecopterans	Abdominal segs. 1–6 (dorsal), 7–10 (ventral)	Contact female abdomen	No[a]	Brinck 1956b	Brinck 1956b
Meleoma (lacewing insect)	Antenna	Holds female antenna	No[a]	Toschi 1965, Tauber 1969	Tauber 1969
	Frontal horns and cavity	Female licks glandular products			
Rheumatobates (water strider)	Antenna	Holds female antenna	No[a,d]	Silvey 1931	Hungerford 1954
	Leg III	Holds female body			
Sepsidae (flies)	Tibia leg I	Holds base of female wing	No[a]	Hennig 1949	Hennig 1949
Many mayflies	Front legs	Hold female prothorax or mesothorax	No[a]	Edmunds et al. 1976; Speith 1940	Edmunds et al. 1976
Many nematodes	Bursa (elaboration of posterior part of body)	Holds female	No[a,b]	Croll and Wright 1976	Chitwood and Chitwood 1974
Litoria (frog)	Thumb[e]	Holds female	No[a]	Noble 1931	Tyler 1976
Some Hyla (frog)	Thumb[e]	Holds female	No[a]	Noble 1931	Duellman 1970
Some Hoplophryne (frog)	Thumb[e]	Holds female	?	Noble 1931	Noble 1931
Leptodactylus (frog)	Thumb, chest[e]	Hold female	No[a]	Heyer 1970	Heyer 1970
Heleophryne (frog)	Fingers, axilla, jaw[e]	Hold female	No[a]	Passmore and Carruthers 1979	Passmore and Carruthers 1979
Some Bufo (toad)	Thumb[e]	Holds female	No[a]	Inger and Greenberg 1956	Inger and Greenberg 1956

a. Lack of female differences deduced from lack of inclusion of the female structure in list of taxonomically useful characters.

b. Author specifically states that females are often indistinguishable without males or that female structures are uniform.

c. Species in this genus that do not "wrap" female antennae during courtship (and thus constitute a control group) all lack the species-specific male antennal modifications.

d. Drawings show general lack of differences between females.

e. Male structures may be used as weapons against other males and may have evolved in male-male combat instead of female choice situations.

Figure 11.1 Sminthurid collembolan antennal grasping organs. *Upper left,* the male seizes his mate with his antennae. *Lower left,* the grasping areas of the male antennae are often complex, while the parts of the female antennae that are grasped are relatively featureless and uniform. *Upper right,* the form of the male antenna is a useful taxonomic character, as shown by partially simplified drawings, each of a different genus. *Lower right,* the detailed drawing of *Jeannenotia* shows more fully the complex structures in the grasping area. (*Upper left,* after Mayer 1967; all others from Massoud and Betsch 1972.)

Nongenitalic male contact structures are especially useful for analyzing the process of rapid and divergent evolution because they contact specific external regions of the female body, and it is thus relatively easy to assess how they are wielded and whether the corresponding female structures vary more or less than those of the males. If these male structures represent an unbiased example of the type of evolution that is typical of genitalia, they can teach us a lot about how and why rapid and divergent genitalic evolution occurs.

An overwhelming trend in Table 11.1, with a few clear exceptions (see Fig. 11.2), is for female structures to be relatively uniform from species to species, despite the sometimes radical changes in the corresponding male structures. This suggests that females of most groups are selecting males on the basis of stimuli rather than mechanical fit. This trend also eliminates the lock and key, mechanical conflict of interest, and pleiotropism (Arnold's modification) hypotheses, all of which predict concomitant male and female changes (see Chapters 2–4). The general absence of female structures that could be said to resist those of the males (see Fig. 11.2) is additional evidence against the mechanical conflict of interest hypothesis.

Some of these nongenitalic structures also give relatively clear evidence of the relative unimportance of sexual selection by direct male-male competition, a point that was left partly unresolved in Chapter 6. Direct male-male competition seems to be ruled out as an important selective factor on male morphology in groups in which females grasp species-specific male structures using their own relatively uniform mouthparts, such as *Argyrodes* and erigonine spiders, *Chordeuma* millipeds, *Meleoma* lacewing insects, *Ectobius* roaches, *Collops* beetles, schizomid arachnids (Figs. 11.3 and 11.4), and in groups in which the "grasping" male structures are physically weak, such as the flexible hairs on the legs of *Rheumatobates* water striders (Fig. 11.5), the antennae of *Belanogaster* wasps and meloid beetles, the tibiae of *Crabro* wasps, the hairs and bristles on the hysterosomes of *Trouessartia* feather mites. Even some of the more powerful clasper organs seem unlikely to function in direct male-male competition. For instance, the clamp on leg II of the spider *Euagrus* (Fig. 11.6), which is used only when the male is briefly inside the female's burrow, is more likely to be important in keeping the female on her back and unable to attack him during copulation and as he attempts to escape immediately afterward than it is in holding on to the female against advances by other males.

An additional aspect of the evolution of nongenitalic contact organs is also revealing. In most (perhaps nearly all) animals, copulation is normally accompanied by female contact with some other parts of the male's body besides his genitalia. Yet many of these other male parts are neither modified nor species-specific in form.

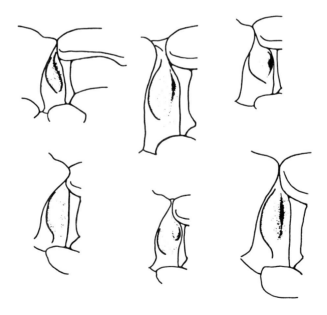

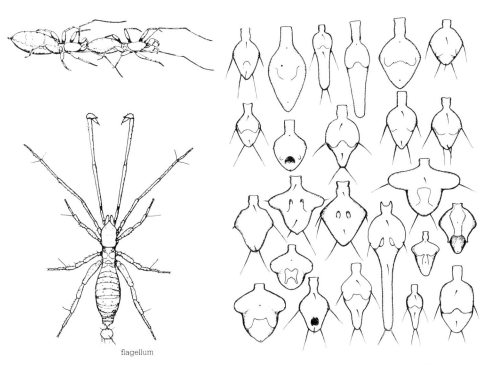

flagellum

Figure 11.3 Upper left, in schizomid arachnids the female (on the left) grasps a special plate, or flagellum on the male's rear end. *Lower left,* the male flagellum is species-specific in form. *Right,* each of the male flagella (bristles are omitted) is from a different species of *Schizomus.* (*Left,* after Strum, and McDonald and Hogue from Kaestner 1968, reprinted by permission of John Wiley & Sons; *right,* after Rowland and Reddell 1979.)

In other words, male structures that are not modified in some way to consistently hold or contact females do not evolve rapidly and divergently, while those that *are* modified for this function show clear signs of rapid and divergent evolution (Fig. 11.7). To explain this, one must answer three different questions. Why does rapid divergence occur so consistently in male structures once they have been modified to contact females? Why do many structures that

---

Figure 11.2 Left, above, the female *Cicindella* tiger beetle has grooves in the side of her thorax where the male's mandibles fit when he mounts her prior to copulation. *Below,* the female's grooves and the apices of the males' mandibles vary in correlated ways between the six species shown (Freitag 1974 and personal communication). As in Fig. 5.4, the species-specific female modifications seem designed to aid the male in clinging to her. In some species the male rides on the female for an extended time. (From Freitag 1974.)

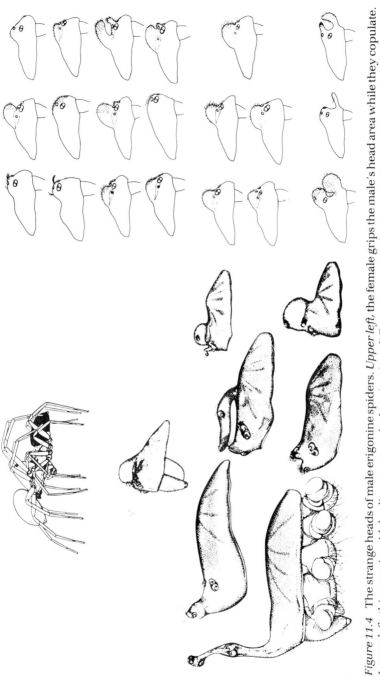

Figure 11.4 The strange heads of male erigonine spiders. *Upper left,* the female grips the male's head area while they copulate. *Lower left,* this area is widely divergent; each drawing is a different genus. *Right,* there is also variation within genera, as illustrated by the profiles of male *Scotinotylus.* (*Left,* from Bristowe 1958; *right,* from Millidge 1981a.)

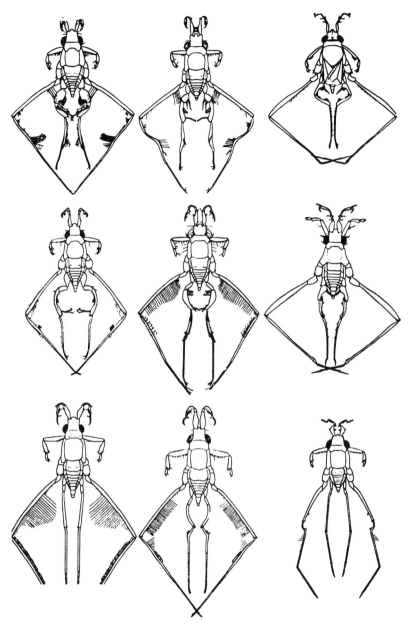

Figure 11.5 Bizarre male antennal and leg modifications (many are flexible hairs) of different species in the water strider genus *Rheumatobates*. Males ride on the backs of females and hold on with their legs and antennae. Female body form varies little between species. (From Hungerford 1959.)

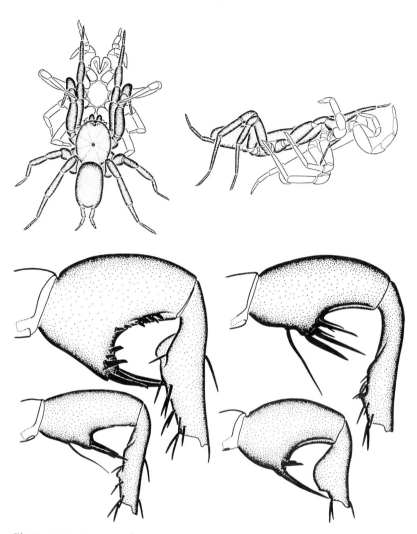

Figure 11.6 Nongenitalic clasper organs of the tarantuloid spider *Euagrus*. *Above,* just before initiating copulation the male (darker) pins the female on her back, using his second legs to hold her second legs. Following copulation the male releases the female and quickly leaves; sometimes the female captures him and kills him. The morphology of the female second leg is quite uniform between species, but that of the male second leg *(below)* differs sharply in different species. (From Coyle in press.)

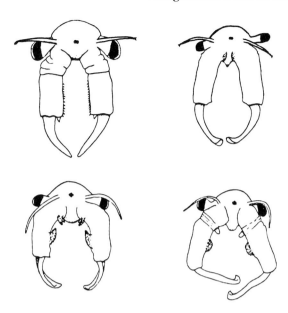

Figure 11.7   Male antennae of four species of the fairy shrimp *Brachinecta*, showing modifications used to hold females. (From Pennak 1978, reprinted by permission of John Wiley & Sons.)

contact females fail to become modified for this function? And why does lack of modification result in lack of rapid and divergent evolution?

The probable explanations are related, first to the inexorable nature of sexual selection by female choice once discrimination arises and, second, to the arbitrary nature of the criteria used by females in discriminating among males. With reference to the first point, the initial modification of a male organ marks the time when selection on males or females (either natural or sexual selection) begins to operate on the male structure to improve its performance in holding, stimulating, and so on. At exactly this same time it begins to become advantageous for females to discriminate among males to insure that their sons have the preferred trait. Thus selection on females to discriminate on the basis of such traits arises more or less simultaneously with the first modification of the organ for sex-associated roles. To take a concrete example, consider the male wings of the flightless mecopteran insect *Boreus.* When the wings first began to be used to grasp females and to acquire their present tonglike configuration (Fig. 12.1), selection on females would have promoted discrimination favoring males with more effective grasping organs. Subsequent male modifications to accentuate those stimuli used by the females, plus changes in the females' criteria, could result in a bout of runaway sexual selection, producing a situation like that found in *B. notoperates*, in which the wings not only clasp the female but are also used to "maul" her antennae and rostrum (Cooper 1974).

The ultimate answer to the question of why females use some cues rather than others is probably determined by interactions among several different factors. On the one hand it is clear from classical ethological studies of "sign stimuli" that in many cases animals respond to only one or a few aspects of the multitude of stimuli they receive (Alcock 1975). Thus we would expect females to perceive and respond to only a few aspects of their mates. In addition, selection in other contexts, for example the walking function of a leg, could act as a counterbalance and make it difficult for runaway selection to occur. Also the likelihood that male variants in any particular structure will arise is highly variable (Bateson 1894), and in some cases a given variant may not yet have arisen; this is the apparent explanation for some patterns of change in courtship structures (see West-Eberhard, 1984). And finally, even male structures that are used to stimulate females would remain unaltered if female discrimination hinged on the male's behavior rather than his morphology. These are admittedly ad hoc arguments in that they do not explain any given case, but our present ignorance of the factors involved makes it unreasonable to expect any other type of explanation: they do, however, offer logical hypothetical explanations for the general overall pattern that is observed.

In summary, the prediction of rapid and divergent evolution in nongenitalic male organs that are specialized to contact females is amply fulfilled. Females apparently are more likely to use the stimulatory rather than the mechanical properties of such organs as cues. Three of the other hypotheses explaining genitalic evolution clearly cannot explain the evolution of these structures. Their evolution illustrates the erratic and arbitrary nature of changes expected under sexual selection by female choice.

12 Conclusions and an overview

In a sense, this book has been one long argument. We have seen that rapid and divergent genitalic evolution is a central evolutionary phenomenon in animals with internal fertilization and that it is part of the even wider trend toward rapid and divergent evolution of any male structure, genitalic or not, that is specialized to contact females in sexual contexts. Because of the pervasiveness of the phenomenon, I believe that it must be based on some very general factor or factors, and I have rejected hypotheses that are valid in only some cases. It seems very unlikely that a combination of different explanations, each with only sporadic validity, is sufficient to explain the extraordinarily consistent evolutionary pattern. There are strong reasons, summarized below, to doubt the validity of the existing hypotheses that have tried to explain rapid and divergent evolution in genitalia.

Lock and key. This hypothesis, which supposes that complex, species-specific genitalic morphology evolves to keep females from being fertilized by males of other species, *(a)* fails to account for the large number of groups in which female genitalia are mechanically incapable of excluding those of nonconspecific males and in which male genitalia are species-specific in form (planarians, nematodes, snails and slugs, cephalopods, snakes, sharks and rays, chaetognaths, some annelids, some mammals, and several other groups). Also, *(b)* female genitalic morphology is often quite invariable in numerous groups in which male structures vary dramatically; less often, male structures are more constant than those of females. In both cases the mutual structural adjustments, which are implicit in the lock and key hypothesis, do not occur. Another reason to doubt this hypothesis is *(c)* that the contexts in which natural selection would favor species recognition by genitalic cues are rare in many groups that have species-specific genitalia. And *(d)* species that have apparently evolved in strict isolation from all close relatives (some parasites and island endemics) and thus probably have never formed cross-specific pairs nevertheless often have complex, species-specific genitalia. Finally, *(e)* the elaborate spermatophores of some terrestrial arthropods, which are as complex and distinct morphologically as many genitalia, are present in those groups of spermatophore-producing species in which cross-specific pairings are least probable.

Genitalic recognition. This hypothesis, which supposes that stimulatory rather than mechanical properties of male genitalia are used by females to identify conspecific males, is contradicted by observations *b, c,* and *d,* listed above.

Pleiotropism. An originally formulated (Mayr 1963), this hypothesis claimed that species-specific differences in genitalic structure are selectively neutral, and that genitalia change as a result of being affected pleiotropically by genes that influence other, selectively important traits. This hypothesis fails to explain *(f)* why genitalia rather than other organs are usually affected pleiotropically; *(g)* why primary genitalia consistently tend not to evolve rapidly and divergently when secondary genitalia are used in sperm transfer; and why in just these cases the secondary genitalic structures (palps, tarsi, and so on) do evolve in this manner. Nor does it explain *(h)* the complete lack of rapid evolutionary divergence in the genitalia of species with external fertilization. Arnold (1973) proposed a modification of the pleiotropism hypothesis that avoids these otherwise fatal weaknesses by abandoning the supposed neutrality of genitalic characters for a lock and key type of argument. His model explained why intromittent structures rather than others should be especially prone to pleiotropic adjustments, but failed to account for observations *a* and *b,* above. In addition, *(i)* there is reason to doubt that the compensatory male and female changes he envisioned are likely to occur often enough to account for the phenomenon.

Mechanical conflict of interest. This hypothesis supposes that males and females engage in coevolutionary races to control the events associated with copulation. It fails to account for frequent disparities in the rates of male and female genitalic evolution, *b* above. In addition, *(j)* it is difficult to accept this hypothesis in the many cases in which species-specific aspects of male genitalia and of genitalic products such as spermatophores are apparently non-manipulative in form. An important point is that *(k)* the lack of obvious anticlasper devices in females does not fit the prediction. Also *(l)* there is reason to suppose that selection often favors females not overcoming male manipulations.

Sexual selection by male conflict. A new hypothesis is that sexual selection in the form of direct male-male competition is responsible for rapid divergence in genitalic structure. I reject this hypothesis because even though such competition is a feasible explanation in some groups (for example, sperm displacement in odonates), *(m)* it is precluded in many other groups by the morphology of male organs, the ways these organs mesh with female organs, and a number of other factors.

Another reason to doubt the previous explanations of genitalic evolution is that nongenitalic male structures that are specialized to contact females in sexual contexts are species-specific and often extraordinarily elaborate in form in a wide variety of animals. The selective factors responsible for this pattern of evolution are probably the same as those involving genitalia; points *a, b, f, j, k,* and *m,* mentioned above, are especially clear with respect to nongenitalic contact structures, giving especially strong reasons to reject all five hypotheses.

Since rapid genitalic divergence is very widespread, and since each of the previous hypotheses is unable to account for substantial numbers of cases, all of the hypotheses are rejected as general explanations. This is not to say that none of these other hypotheses is ever true. More than one selective factor may act on the evolution of genitalia, and they could act in a variety of combinations with each other. An explanation does not have to work for all cases in order to be true for specific ones. Probably each of the hypotheses discussed in this book is true for at least a few cases; after all, we are dealing with perhaps several million species.

My hypothesis, sexual selection by female choice, proposes that male genitalia function as "internal courtship" devices to increase the likelihood that females will actually use a given male's sperm to fertilize her eggs rather than those of another male. According to this hypothesis, the diversity of male genitalic form is the result of runaway evolutionary processes generated by sexual selection by female choice on otherwise arbitrary aspects of male genitalia. This hypothesis is compatible with the data and with arguments *a* to *m* given above. In addition it explains the dominant pattern, heretofore unremarked, that it is the male rather than the female that introduces gametes into the partner's body in species with internal fertilization. The only documented exception to this pattern, in seahorses and their relatives, involves, as predicted by the sexual selection hypothesis, species with unusually high paternal investment in offspring. Sexual selection by female choice also explains the trend toward rapid divergence of nongenitalic male organs that are specialized to contact females prior to or during copulation. It also explains the apparently greater conservatism in the morphology of female as opposed to male genitalia; the "extravagance" and apparent lack of utility of many male genitalic structures; the apparently stimulatory functions, during copulation, of the male genitalia of a wide variety of species; the probable association between species specificity and complexity of male genitalia; and the possible association between elaborate premating courtship display and simpler, more uniform genitalia.

There are several apparent drawbacks to the female choice hypothesis. The hypothesis requires that in groups with species-specific genitalia a single copulation does not always result in fertilization of the female's entire complement of eggs. The assumption that copulation necessarily entails fertilization turns out to be a misconception, since a number of female-mediated processes are generally necessary to insure the use of a given male's sperm in fertilization. "Cryptic" female choice is feasible in most animal groups. Some species' extraordinarily tortuous morphology in those parts of the female reproductive tract that are traversed by sperm, and the "reinvention" of sperm-conducting female reproductive tracts (paragenitalia) in bedbugs and their relatives, are striking examples of the possible importance to females of being able to exercise such choice: the evidence strongly suggests, in fact, that bedbug paragenitalia evolved as sperm-*killing* and *de*activating organs!

Another difficulty with the hypothesis is that it can work only if individual females not infrequently make genitalic contact with more than a single male during their reproductive lifetimes. This runs counter to the idea that in many species females mate only once. But the frequency of remating by females in the field has not been convincingly documented in most groups. What data do exist suggest that females generally do sometimes remate, except in a few groups like termites, where, as predicted, genitalia are simple and not species-specific in form. The most complete remating data come from spermatophore counts in the reproductive tracts of female lepidopterans; in this group female remating is very widespread.

A third major objection is that even if females do sometimes remate, they will mate with relatively few males. They will thus have fewer males to choose among than females of the species classically associated with sexual selection by female choice, such as those that mate in leks. These moderate intensities of selection might not be enough to produce the rapid changes observed in genitalia. But a lower rate of genitalic change may persist over a longer period of time than in classical cases of sexual selection by female choice, such as bird plumage and frog calls; details of genitalic structures are seldom likely to be associated with additional disadvantages, such as increased risk of predation, so runaway processes involving genitalia are much less likely to be halted. It is thus feasible for such processes to bring about rapid evolutionary changes.

The evidence suggests that sexual selection by female choice is the only hypothesis that can explain the great majority of the cases, and thus it is the most likely to account for the extraordinary generality of the phenomenon. Tests that could either disprove or modify the hypothesis could be made in a number of areas. The frequency of female remating in the field and correlations (or lack of correlation) with genitalic diversity can be tested in other groups as was done for *Heliconius* butterflies in Chapter 8. The neuroanatomy and physiology of genitalia in species in which it seems likely that females use sensory criteria to discriminate among males could clarify females' abilities to make discriminations; such studies are especially feasible in species that have species-specific male grasping organs, both genital and nongenital, that contact the outside of the female's body, because the corresponding female structures are relatively accessible (trailbreaking studies in this area are those of Loibl 1958; Krieger and Krieger-Loibl 1958; and Robertson and Paterson 1982). Breeding experiments can be done using genetic markers linked with male genitalic variants to determine whether females can discriminate minor differences in males' genitalia, as predicted. Genitalia can be carefully measured for (1) character displacement and (2) differing degrees of variability in populations that are sympatric with close relatives as compared with those that are isolated. The movements of species-specific male organs can be studied while on the female, when the male is lightly anesthetized (West-Eberhard 1984), and perhaps also, using X-ray movies, while inside the female. Sperm transfer during copulation can be timed to check for postinsemination courtship. Studying the effects of experimental modifications of species-specific male organs may be especially feasible in species that are both relatively large and easy to keep in captivity; poeciliid fish come to mind as candidates.

An understanding of genitalic evolution illuminates the very strong association between internal fertilization and the tendency for females rather than males to provide parental care to offspring. The reason for this trend is probably the fact that fertilization consistently occurs within the female rather than the male (Gross and Shine 1981). The arguments in Chapter 5 regarding the evolution of male versus female intromittent organs form an important part of the explanation of the pattern of male and female differences in parental investment throughout the animal kingdom.

The analysis of genitalic evolution as a result of sexual selection also dramatizes an aspect of the theory itself that is sometimes overlooked. Discrimination among males on the basis of their genitalia could seldom result in either material benefits to the female or superiority in her offspring's ability to grow and survive. Indeed, a male's genitalia are probably relatively poor indicators of his overall fitness. Nevertheless, because of their almost inevitable physical and temporal association with female processes that influence male reproduction (mechanical fit, sperm transport, and so on), male genitalia can influence the probability that females will execute these crucial reproductive processes. In general, once any male signal is used by females in ways that affect the male's reproductive success, selection can favor both male enhancement of the signal's effectiveness and female discrimination on the basis of the signal itself, even though the signal is not correlated with any other aspects of the organism's fitness. It is as if, because of a quirk in the way selection works, an entire array of ever-changing and otherwise superfluous, extravagant characters is carried along through evolutionary time by all of the "useful" characters that enable the organism to find appropriate habitats, acquire and metabolize food, defend itself from predators and parasites, and so on.

Another important point that the hypothesis brings into focus is the arbitrariness and inadequacy of the term *genitalia*. As in many instances in evolutionary biology, this insight is really just an extension of Darwin's original ideas; he noted that special male organs for holding females "graduate into those that are commonly ranked as primary, and in some cases can be hardly distinguished from them" (Darwin 1871, p. 567), and he discussed virtually every other category of sexually dimorphic, extravagantly developed courtship-associated character. Biologists usually name structures on the basis of either homology or function. Genitalic structures in different animal groups are obviously not homologous, and the common criterion for classing structures as "genitalic" has been functional association with gamete transfer. But we have seen that many genitalic structures probably have the additional or alternative function of courtship and competitive insemination. Is it reasonable to consider as genitalia species-specific structures such as the *Erebia* butterfly "claspers," which are really used to stroke the female abdomen during copulation, but to insist that the species-specific male wings of the scorpionfly *Boreus* (Fig. 12.1), which seize the female and are then raked across her antennae, are not genitalia? Some structures that have been traditionally called genitalia probably function in courtship, others in gamete transfer, and others (the majority?) in a combination of the two. This book thus has the dubious distinction of questioning usefulness of its own title. As

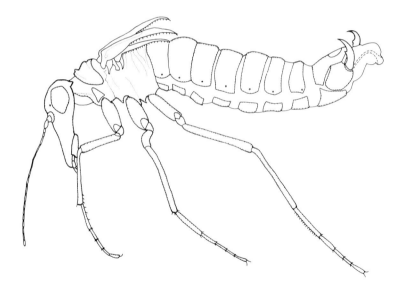

Figure 12.1 A male *Boreus* scorpionfly, showing his highly modified tong-like wings, used to seize the female and then rake across her antennae prior to copulation. The wings are often species-specific in form, and their seizing and apparently stimulatory functions are similar to those of many abdominal structures that are usually classified as genitalia. (After Cooper 1974.)

with other biological dichotomies, such as plant versus animal, our language is misleading, and the natural world has proven to be more complex than is suggested by linguistic distinctions.

I suggest that genitalic evolution fits into the even larger evolutionary tendency toward rapid divergence of signals that are used in intraspecific competition, whether for mates or other resources (West-Eberhard 1983). Thus in a very real sense the arguments made in this book are an attempt to bring genitalic evolution, long neglected by those evolutionary biologists outside of taxonomy, into the mainstream of evolutionary theory.

References

Index

References

Aamdal, J., J. Fougner, and K. Nyberg. 1978. Artificial insemination in foxes. *Symposium of the Zoological Society of London* 43:241–248.

Abalos, J. W. 1968. La transferencia espermática en los arácnidos. *Revista de la Universidad Nacional de Córdoba* (Argentina), 2d ser. 9:251–278.

Abalos, J. W., and E. C. Baez. 1966. Las arañas del género *Latrodectus* en Santiago del Estero. *Revista de la Facultad de Ciencias Exactas, Físicas v Naturales de la Universidad Nacional de Cordoba* (Argentina) 27(3–4):1–30.

Abele, L. G., and S. Gilchrist. 1977. Homosexual rape and sexual selection in acanthocephalan worms. *Science* 197:81–83.

Adams, T., and A. Hinz. 1969. Relationship of age, ovarian development, and the corpus allatum to mating in the house-fly, *Musca domestica*. *Journal of Insect Physiology* 15:201–215.

Adams, C. L., and R. B. Selander. 1979. The biology of blister beetles of the vittata groups of the genus *Epicauta* (Coleoptera, Meloidae). *Bulletin of the American Museum of Natural History* 162:137–266.

Adler, N. T. 1969. Effects of the male's copulatory behavior on successful pregnancy of the female rat. *Journal of Comparative and Physiological Psychology* 69:613–622.

Ahmad, I., and S. Sheikh. 1983. Revision of the genus *Anhomoeus* Hsiao 1963 (Hemiptera: Coreidae: Coreinae), with description of two new species. *Annals of the Entomological Society of America* 76:853–859.

Alberti, G. 1974. Fortpflanzungsverhalten und Fortpflanzungsorgane der Schnabelmilven (Acarina: Bdellidae, Trombidiformes). *Zeitschrift für Morphologie Tiere* 78:111–157.

Alcock, J. 1975. *Animal behavior.* Sunderland, Mass.: Sinauer.

―――― 1976. Courtship and mating in *Hippomelas planicosta* (Coleoptera: Buprestidae). *Coleopterist's Bulletin* 30(4):343–348.

―――― 1979. The evolution of intraspecific diversity in male reproductive strategies in some bees and wasps. In *Sexual selection and reproductive competition in insects,* ed. M. Blum and N. Blum. New York: Academic Press.

Alcock, J., C. Jones, and S. Buchmann. 1976. Location before emergence of the female bee *Centris pallida* by its male (Hymenoptera: Anthophoridae). *Journal of Zoology* (London) 179:189–199.

Alcock, J., and S. Buchmann. In press. The significance of post-insemination display by male *Centris pallida* (Hymenoptera: Anthophoridae). *Zeitschrift für Tierpsychologie.*

Alexander, A. J. 1962a. Biology and behavior of *Damon variegatus* Perty of South Africa and *Admetus barbadensis* Pocock of Trinidad, W.I. (Arachnida, Pedipalpi). *Zoologica* 47(1):25–37.

——— 1962b. Courtship and mating in amblypygids (Pedipalpi, Arachnida). *Proceedings of the Zoological Society of London* 138(3):379–383.

Alexander, R. D. 1959. The courtship and copulation of *Pasimachus punctulatus* Haldemann (Coleoptera: Carabidae). *Annals of the Entomological Society of America* 52:485.

——— 1962. The role of behavioral study in cricket classification. *Systematic Zoology* 11(2):53–72.

——— 1964. The evolution of mating behaviour in arthropods. *Symposium of the Royal Entomological Society of London* 2:78–94.

Alexander, R. D. Manuscript. The study of animal communication: A new era.

Alexander, R. D., and T. Moore. 1962. The evolutionary relationships of 17-year and 13-year cicadas, and three new species (Homoptera, Cicadidae, *Magicicada*). *Miscellaneous Publications of the Museum of Zoology* University of Michigan 121:5–59.

Alexander, R. D., and D. Otte. 1967. The evolution of genitalia and mating behavior in crickets (Gryllidae) and other Orthoptera. *Miscellaneous Publications of the Museum of Zoology* University of Michigan 133:1–62.

Allison, A. S. 1975. Sperm transport in the ewe: the effect of ovarian steroids. In *The biology of spermatozoa,* ed. E. S. E. Hafez and C. G. Thibault. New York: S. Karger.

Alvarado, A., J. Arias, D. Briceño, W. Eberhard, R. Lopez, O. Rocha, E. Rojas, and O. Trejos. Manuscript. Interspecific variation, assortative mating, and the lock and key hypothesis for genitalic evolution.

Alvariño, A. 1965. Chaetognaths. *Oceanography and Marine Biology: An Annual Review* 3:115–194.

Ander, K. 1970. Orthoptera Saltatoria. In *Taxonomist's glossary of genitalia of insects,* ed. S. L. Tuxen. Darien, Conn.: S-H Service Agency.

Anderson, J. D. 1961. The courtship behavior of *Ambystoma macrodactylum croceum. Copeia* 1961(2):132–139.

Anderson, N. M. 1981. Semiaquatic bugs: phylogeny and classification of the Hebridae (Heteroptera: Gerromorpha) with revisions of *Timasius, Neotimasius* and *Hyrcanus. Systematic Entomology* 6:377–412.

Andrews, E. A. 1904. Breeding habits of crayfish. *American Naturalist* 38:165–206.

——— 1911. Male organs for spermatophore transfer in the crayfish, *Cambarus affinis:* their structure and use. *Journal of Morphology* 22(2):239–297.

Ansari, M. A. R. 1956. A revision of the *Brüelia* (Mallophaga) species infesting the Corvidae. *Bulletin of the British Museum of Natural History (Entomology)* 4:371–406.

——— 1957. A revision of the *Brüelia* (Mallophaga) species infesting the Corvidae, II. *Bulletin of the British Museum of Natural History (Entomology)* 5:145–182.

Applegate, S. P. 1967. A survey of shark hard parts. In *Sharks, skates and rays*, ed. P. Gilbert, R. Matthewson, and D. Rall. Baltimore: Johns Hopkins University Press.

Archer, A. F. 1951a. Studies in the orbweaving spiders (Argiopidae). *American Museum Novitates* 1487:1–52.

——— 1951b. Studies in the orbweaving spiders (Argiopidae). *American Museum Novitates* 1502:1–334.

Arikawa, K., E. Eguchi, A. Yoshida, and K. Aoki. 1980. Multiple extraocular photoreceptive areas on genitalia of butterfly *Papilio xanthus*. *Nature* 288:700–702.

Arnold, E. N. 1973. Relationships of the palaearctic lizards assigned to the genera *Lacerta, Algyroides* and *Psammordromus* (Reptilia: Lacertidae). *Bulletin of the British Museum of Natural History (Zoology)* 25(8):291–366.

Arnold, J. M., and L. D. Williams-Arnold. 1977. Cephalopoda: Decapoda. In *Reproduction of marine invertebrates*, vol. 4, ed. A. Giese and J. Pearse. New York: Academic Press.

Arnold, S. J. 1972. The evolution of courtship behavior in salamanders. Ph.D. diss., University of Michigan.

Askew, R. 1968. Considerations on speciation in Chalcidoidea (Hymenoptera). *Evolution* 22(3):642–645.

Austad, S. N. 1982. First male sperm priority in the bowl and doily spider, *Frontinella pyramita* (Walckenaer). *Evolution* 36(4):777–785.

Avel, M. 1959. Classe des annelides oligochetes. In *Traité de zoologie*, vol. 5, ed. P. P. Grassé. Paris: Masson.

Bacheler, J. S., and D. H. Habeck. 1974. Biology and hybridization of *Apantesis phalerata* and *A. radians* (Lepidoptera: Arctiidae). *Annals of the Entomological Society of America* 67(6):971–975.

Baker, F. C. 1945. *The molluscan family Planorbidae*. Urbana: University of Illinois Press.

Baldwin, F. T., and E. H. Bryant. 1981. Effect of size upon mating preference within geographic strains of the house fly, *Musca domestica*. *Evolution* 35:1134–1141.

Ballantyne, L., and M. McLean. 1970. Revisional studies on the firefly genus *Pteroptyx* Olivier (Coleoptera: Lampyridae: Luciolinae: Luciolini). *Transactions of the American Entomological Society* 96:223–305.

Bänninger, M. 1950. The subtribe Pasimachina (Coleoptera, Carabidae, Scaritini). *Revista Entomologica Rio de Janeiro* 21(3):481–511.

Bateson, W. 1894. *Materials for the study of variation.* London: Macmillan.

Batra, S. W. T. 1978. Aggression, territoriality, mating and nest aggregation of some solitary bees (Hymenoptera): Halictidae, Megachilidae, Colletidae, Anthophoridae). *Journal of the Kansas Entomological Society* 51(4):547–559.

Beeman, R. D. 1977. Gasteropoda: Opisthobranchia. In *Reproduction of marine invertebrates,* vol. 4, ed. A. Giese and J. Pearse. New York: Academic Press.

Beier, M. 1970. Dictyoptera. In *Taxonomist's glossary of genitalia in insects,* ed. S. L. Tuxen. Darien, Conn.: S-H Service Agency.

Belding, D. L. 1958. *Basic clinical parasitology.* New York: Appleton-Century-Crofts.

Bellomy, M. D. 1969. *Encyclopedia of sea horses.* Jersey City, N.J.: T. F. H. Publishers.

Benz, G. 1969. Influence of mating, insemination, and other factors on oogenesis and oviposition in the moth *Zeiraphera diniana. Journal of Insect Physiology* 15:55–71.

Berlese, A. 1925. *Gli Insetti,* vol. 2. Milan: Societa Editrice Libraria.

Berry, A. J. 1977. Gasteropoda: Pulmonata. In *Reproduction of marine invertebrates,* vol. 4, ed. A. Giese and J. Pearse. New York: Academic Press.

Beveridge, I. 1982. A taxonomic revision of the Pharyngostrongylinae Popova (Nematoda: Strongyloidea) from macropodid marsupials. *Australian Journal of Zoology* (Suppl. Ser.) 83:1–150.

Bick, G., and J. Bick. 1981. Heterospecific pairing among Odonata. *Odonatologica* 10(4):259–270.

Bieman, D. N., and J. A. Witter. 1982. Mating wounds in *Malacosoma:* an insight into bedbug mating behavior. *Florida Entomologist* 65:377–378.

Birdsall, D. A., and D. Nash. 1973. Occurrence of successful multiple insemination of females in natural populations of deer mice *(Peromyscus maniculatus). Evolution* 27:106–110.

Bishop, D. W. 1971. Sperm transport in the Fallopian tube. In *Pathways to conception,* ed. A. Sherman. Springfield, Ill.: Thomas.

Bitsch, J. 1979. Morphologie abdominale des Insectes. In *Traité de zoologie,* vol. 8, fasc. 2, ed. P. P. Grassé. Paris: Masson.

Blandau, R. J. 1969. Gamete transport—comparative aspects. In *The mammalian oviduct,* ed. E. S. E. Hafez and R. J. Blandau. Chicago: University of Chicago Press.

———— 1973. Sperm transport through the mammalian cervix: comparative aspects. In *The biology of the cervix,* ed. R. J. Blandau and K. Moghissi. Chicago: University of Chicago Press.

Blest, A. D., and G. Pomeroy. 1978. The sexual behavior and genital mechanics of three species of *Mynoglenes* (Araneae: Linyphiidae). *Journal of Zoology* (London) 185:319–340.

Blum, M., and N. Blum, eds. 1979. *Sexual selection and reproductive competition in insects.* New York: Academic Press.

Blumer, L. 1979. Male parental care in the bony fishes. *Quarterly Review of Biology* 54:149–161.

Boggs, C. L. 1981. Selection pressures affecting male nutrient investment at mating in heliconiine butterflies. *Evolution* 35(5):931–940.

Bohart, R. M., and A. Menke. 1976. *Sphecid wasps of the world.* Los Angeles: University of California Press.

Bohart, R. M., and L. A. Stange. 1965. A revision of the genus *Zethus* Fabricius in the Western Hemisphere. *University of California Publications in Entomology* 40:1–208.

Bohme, W. 1983. The Tucano indians of Colombia and the iguanid lizard *Plica plica:* ethnological, herpetological and ethological implications. *Biotropica* 15:148–150.

Boisseau, C., and J. Joly. 1975. Transport and survival of spermatozoa in female Amphibia. In *The biology of spermatozoa,* ed. E. S. E. Hafez and C. G. Thibault. New York: S. Karger.

Borgia, G. 1979. Sexual selection and the evolution of mating systems. In *Sexual selection and reproductive competition in insects,* ed. M. Blum and N. Blum. New York: Academic Press.

Böttger, K. 1962. Zur Biologie und Ethologie der einheimischen Wassermilben *Arrenurus (Megaluracarus) globator* (Mull.), 1776 *Piona nodata* (Mull.), 1776 und *Eylais infundibulifera meridionalis* (Thon), 1899 (Hydrachnellae, Acari). *Zoologische Jahrbücher Abteilung für Systematik* 89:501–584.

——— 1965. Zur Ökologie und Fortpflanzungsbiologie von *Arrenurus valdiviensis* K. O. Viets 1964 (Hydrachnellae, Acari). *Zeitschrift für Morphologie und Ökologie der Tiere* 55:115–141.

Boulard, M. 1965. L'appareil génital ectodermique des cigales femelles *Annales Societé Entomologique de France* 1:797–812.

Bousfield, E. L. 1958. Fresh-water amphipod crustaceans of glaciated North America. *Canadian Field-Naturalist* 72(2):55–113.

Branch, W. R. In press. Hemipeneal morphology in platynotan lizards. *Journal of Herpetology.*

Breder, C. M., and D. E. Rosen. 1966. *Modes of reproduction in fishes.* Garden City, N.Y.: Natural History Press.

Briegleb, W. 1961. Die Spermatophore des Grottenolms. *Zoologischer Anzeiger* 166(1–2):87–91.

Brignoli, P. M. 1974. On some Ricinulei of Mexico with notes of the morphology of the female genital apparatus (Arachnida, Ricinulei). *Accademia Nazionale dei Lincei* Quaderno No. 171:153–174.

——— 1978. Some remarks on the relationships between the Haplogynae, the Semientelegynae and the Cribellatae (Araneae). *Symposium of the Zoological Society of London* 42:285–292.

Brinck, R. 1956a. Reproductive system and mating in Plecoptera, I. *Opuscula Entomologica.* 21(1):57–96.

——— 1956b. Reproductive system and mating in Plecoptera, II. *Opuscula Entomologica.* 21(2–3):97–128.

Brindle, A. 1971. Bredin-Archbold-Smithsonian biological survey of Dominica. The Dermaptera (earwigs) of Dominica. *Smithsonian Contributions to Zoology* 63:1–25.

Bristowe, W. S. 1958. *The world of spiders.* London: Collins.

Brooks, D. R., and D. R. Glen. 1982. Pinworms and primates: a case study in coevolution. *Proceedings of the Helminthological Society of Washington* 49:76–85.

Brown, K. S. 1981. The biology of *Heliconius* and related genera. *Annual Review of Entomology* 26:427–456.

Brown, K. S., Jr., P. M. Sheppard, and J. R. G. Turner. 1974. Quaternary refugia in tropical America: evidence from race formation in *Heliconius* butterflies. *Proceedings of the Royal Society of London* (B) 187:369–378.

Brown, W. L., and E. O. Wilson. 1956. Character displacement. *Systematic Zoology* 5:49–64.

Burns, J. M. 1968. Mating frequency in natural populations of skippers and butterflies as determined by spermatophore counts. *Proceedings of the National Academy of Sciences* (U.S.A.) 61:852–859.

———— 1970. Secondary symmetry of asymmetric genitalia in males of *Erynnis funeralis* and *E. propertius* (Lepidoptera: Hesperiidae). *Psyche* 77:430–435.

Burt, D. R. R. 1970. Platyhelminthes and parasitism. New York: Elsevier.

Busck, A. 1931. On the female genitalia of the microlepidoptera and their importance in the classification and determination of these moths. *Bulletin of the Brooklyn Entomological Society* 26(5):199–216.

Bush, A. O., and W. M. Samuel. 1978. The genus *Travassosius* Khalil, 1922 (Nematoda, Trichostrongyloidea) in beaver, *Castor* spp.: a review and suggestion for speciation. *Canadian Journal of Zoology* 56:1471–1474.

Byers, G. W. 1970. New and little known Chinese Mecoptera. *Journal of the Kansas Entomological Society* 43(4):383–394.

Cade, W. 1979a. Effect of male-deprivation on female phonotaxus in field crickets (Orthoptera: Gryllidae: *Gryllus*). *Canadian Entomologist* 111:741–744.

———— 1979b. The evolution of alternative male reproductive strategies in field crickets. In *Sexual selection and reproductive competition in insects,* ed. M. Blum and N. Blum. New York: Academic Press.

Callahan, P. S., and J. B. Chapin. 1960. Morphology of the reproductive system and mating in two representative members of the family Noctuidae, *Pseudaletia unipuncta* and *Peridroma margaritosa,* with comparison to *Heliothis zea. Annals of the Entomological Society of America* 53:763–782.

Campbell, I. M. 1961. Polygyny in *Choristoneura* Led. (Lepidoptera: Tortricidae). *Canadian Entomologist* 93:1160–1162.

Carayon, J. 1966. Traumatic insemination and the paragenital system. In *Monograph of Cimicidae,* ed. R. Usinger. Philadelphia: Thomas Say Foundation 7, Entomological Society of America.

—— 1975. Insémination extra-génitale traumatique. In *Traité de zoologie,* vol. 8, ed. P. P. Grassé. Paris: Masson.

Carvajal, J. 1977. Description of the adult and larva of *Caulobothrium myliobatidis* sp. n. (Cestoda: Tetraphyllidae) from Chile. *Journal of Parasitology* 63:99–103.

Carvalho, J. C. M., and W. C. Gagne. 1968. Miridae of the Galapagos Islands (Heteroptera). *Proceedings of the California Academy of Sciences,* 4th ser. 36:147–219.

Chamberlin, R. V., and W. Ivie. 1943. A new genus of theridiid spiders in which the male develops only one palpus. *Bulletin of the University of Utah* 24:190–220.

Chaplin, S. J. 1973. Reproductive isolation between two sympatric species of *Oncopeltus* (Hemiptera: Lygaeidae) in the tropics. *Annals of the Entomological Society of America* 66(5):997–1000.

Chapman, R. 1969. *The insects, structure and function.* London: English University Press.

Chester, R. V., and I. Zucker. 1970. Influence of male copulatory behavior on sperm transport, pregnancy and pseudopregnancy in female rats. *Physiological Behavior* 5:35–43.

Chickering, A. M. 1959. The genus *Tetragnatha* (Araneae, Argiopidae) in Michigan. *Bulletin of the Museum of Comparative Zoology* 119(9):475–499.

Chitwood, B. G., and M. B. Chitwood. 1974. *Introduction to nematology.* Baltimore: University Park Press.

Chvala, M. 1971. Redescriptions of *Hilara* species described by G. Strobl from Sapin (Diptera, Empididae). *Acta Entomologica Bohemoslovaca* 68:322–340.

Chvala, M., J. Droskocil, J. H. Mook, and V. Podorny. 1974. The genus *Lipara* Meigen (Diptera, Chloropidae), systematics, morphology, behaviour, and ecology. *Tijdschrift voor Entomologie* 117:1–25.

Claridge, M. F., and W. J. Reynolds. 1972. Host plant specificity, oviposition behaviour and egg parasitism in some woodland leafhoppers of the genus *Oncopsis* (Hemiptera: Homoptera: Cicadellidae). *Transactions of the Royal Entomological Society of London* 124(2):149–166.

—— 1973. Male courtship songs and sibling species in the *Oncopsis flavicollis* species group (Hemiptera: Cicadellidae). *Journal of Entomology* (B) 42(1):29–39.

Clark, E., and L. R. Aronson. 1951. Sexual behavior in the guppy *Lebistes reticulatus* (Peters). *Zoologica* 36(1):49–66.

Clark, E., L. R. Aronson, and M. Gordon. 1954. Mating behavior patterns in two sympatric species of xiphophorin fishes: their inheritance and significance in sexual isolation. *Bulletin of the American Museum of Natural History* 103(2):139–225.

Clay, T. 1974. The Phthiraptera (Insecta) parasitic on flamingos (Phoenicopteridae: Aves). *Journal of Zoology* (London) 172:483–490.

—— 1976. The species of *Ibidoecus* (Phthiraptera) on *Threskiornis* (Aves). *Systematic Entomology* 1:1–7.

Clements, A. 1963. *The Physiology of Mosquitoes.* New York: Macmillan.

Clough, G. 1969. Some preliminary observations on reproduction in the warthog, *Phacochoerus aethiopicus* Pallas. *Journal of Reproduction and Fertility Supplement* 6:323–337.

Cobbs, G. 1977. Multiple insemination and male sexual selection in natural populations of *Drosophila pseudoobscura. American Naturalist* 111:641–656.

Coe, M. J. 1969. The anatomy of the reproductive tract and breeding in the spring haas, *Pedetes surdaster larvalis* Hollister. *Journal of Reproduction and Fertility Supplement* 6:159–174.

Coiffait, H. 1981. Contribution à la connaissance des staphylinidés des Iles Galapagos (Coleoptera). *Annales de la Societé entomologique de France,* n.s. 17:287–310.

Coineau, Y. 1976. Les pariades sexuelles des Saxidrominae Coineau 1974 (Acariens Prostigmates, Adamystidae). *Acarologia* 18:234–240.

Cole, L. J. 1901. Notes on the habits of pycnogonids. *Biological Bulletin* (Woods Hole) 2:195–207.

Collin, J. 1961. *British flies: Empididae.* Cambridge: Cambridge University Press.

Common, I. F. B. 1970. Lepidoptera. In *The insects of Australia.* Carleton, Victoria: Melbourne University Press.

Comstock, J. H. 1967. *The Spider Book,* rev. and ed. W. J. Gertsch. Ithaca, New York: Comstock Publishing Associates.

Coninck, L. de. 1965. Clase des nematodes. In *Traité de Zoologie,* vol. 4, fasc. 1, ed. P. Grassé. Paris: Masson.

Connor, J., and D. Crews. 1980. Sperm transfer and storage in the lizard, *Anolis carolinensis. Journal of Morphology* 163:331–348.

Connor, W. E., T. Eisner, R. K. Van der Meer, A. Guerrero, D. Ghiringelli, and J. Meinwald. 1980. Sex attractant of an arctiid moth *(Utetheisa ornatrix):* a pulsed signal. *Behavioral Ecology and Sociobiology* 7:55–63.

Connor, W. E., T. Eisner, R. K. Van der Meer, A. Guerrero, and J. Meinwald. 1981. Precopulatory sexual interaction in an arctiid moth *(Utetheisa ornatrix):* role of pheromone derived from dietary alkaloids. *Behavioral Ecology and Sociobiology* 9:227–235.

Coope, G. R. 1979. Late cenozoic fossil Coleoptera: evolution, biogeography, and ecology. *Annual Review of Ecology and Systematics* 10:247–267.

Cooper, K. W. 1972. A southern Californian *Boreus, B. notoperates* n. sp. I: Comparative morphology and systematics (Mecoptera: Boreidae). *Psyche* 79(4):269–283.

——— 1974. Sexual biology, chromosomes, development, life histories, and parasites of *Boreus,* especially of *B. notoperates,* a southern California *Boreus.* II. (Mecoptera: Boreidae). *Psyche* 81(1):84–120.

Corbet, P. 1962. *A biology of dragonflies.* London: Witherby.

Cowan, D. in prep. Copulatory behavior of eumenid wasps (Hymenoptera).

Coyle, F. A. 1969. The mygalomorph genus *Atypoides* (Araneae: Antrodiae-tidae). *Psyche* 75(2):157–194.

——— 1971. Systematics and natural history of the mygalomorph spider genus *Antrodiaetus* and related genera (Araneae: Antrodiaetidae). *Bulletin of the Museum of Comparative Zoology* 141(6):269–402.

———— 1974. Systematics of the trapdoor spider genus *Aliatypus* (Araneae: Antrodiaetidae). *Psyche* 81(3–4):431–500.

———— 1981. The mygalomorph spider genus *Microhexura* (Araneae, Dipluridae). *Bulletin of the American Museum of Natural History* 170:64–75.

———— in press. Courtship, mating, and the function of male-specific leg structures in the mygalomorph spider genus *Euagrus* (Araneae, Dipluridae). *Proceedings of the Ninth International Congress of Arachnology* (Panama).

Craig, G. 1967. Mosquitoes: female monogamy induced by male accessory gland substance. *Science* 156:1499–1501.

Crane, J. 1975. *Fiddler crabs of the world (Ocypodidae: genus Uca).* Princeton: Princeton University Press.

Croll, N. A., and K. A. Wright. 1976. Observations on the movements and structure of the bursa of *Nippostrongylus brasiliensis* and *Nematospiroides dubius. Canadian Journal of Zoology* 54:1466–1480.

Cross, E. A. 1965. The generic relationships of the family Pymotidae (Acarina: Trombidiformes). *University of Kansas Science Bulletin* 45(2):29–275.

Crozier, R. H. 1977. Evolutionary genetics of the Hymenoptera. *Annual Review of Entomology* 22:263–288.

Cuellar, O. 1966. Oviductal anatomy and sperm storage structures in lizards. *Journal of Morphology* 119:7–20.

Cunningham, R., G. Farias, S. Nakagawa, and D. Chambers. 1971. Reproduction in the Mediterranean fruit fly: depletion of stored sperm in females. *Annals of the Entomological Society of America* 64(1):312–313.

Cutler, B. 1971. *Darwinneon crypticus,* a new genus and species of jumping spider from the Galapagos Islands (Araneae: Salticidae). *Proceedings of the California Academy of Sciences,* 4th ser. 37:509–513.

Daly, M. 1978. The cost of mating. *American Naturalist* 112:771–774.

Daly, M., and M. Wilson. 1978. *Sex, evolution, and behavior.* North Scituate, Mass.: Duxbury Press.

Darwin, C. 1871. *The descent of man and selection in relation to sex.* Reprinted, New York: Modern Library.

Davey, K. G. 1960. The evolution of spermatophores in insects. *Proceedings of the Royal Entomological Society of London* (A) 35:107–114.

———— 1965. *Reproduction in insects.* London: Oliver and Boyd.

Davis, C. 1949. The pelagic Copepoda of the northeastern Pacific Ocean. *University of Washington Publications in Biology* 14:1–118.

Deboutteville, C. D. 1970. Zoraptera. In *Taxonomist's glossary of genitalia of insects,* ed. S. L. Tuxen. Darien, Conn.: S-H Service Agency.

Deeleman-Reinhold, C., and P. R. Deeleman. 1980. Remarks on troglobitism in spiders. *Proceedings of the Eighth International Arachnological Congress* (Vienna) 433–438.

Degrugillier, M. E., and R. A. Leopold. 1972. Abdominal and peripheral nervous system of the adult female house fly and its role in mating behavior and insemination. *Annals of the Entomological Society of America* 65(3):689–695.

Deitz, L. L. 1975. Classification of the higher categories of the New World treehoppers (Homoptera: Membracidae). *North Carolina Agricultural Station Technical Bulletin* 225:1–130.

DeLong, D. M., and P. H. Freytag. 1972a. Studies of the Gyponinae: the genus *Folicana* and nine new species. *Journal of the Kansas Entomological Society* 45(3):282–295.

―――― 1972b. Studies of the Gyponinae: the genus *Culumana* and seven new species. *Journal of the Kansas Entomological Society* 45(4):405–413.

Denier, P. 1934. Contribución al estudio de los melóidos americanos, II: ensayo de clasificación de las *Pyrota* (Dej.) Lec. basada en los caracteres sexuales secundarios de los machos. *Revista de la Sociedad Entomológica de Argentina* 6(1):49–75.

DeWilde, J. 1964. Reproduction. In *The physiology of Insecta,* ed. M. Rockstein. New York: Academic Press.

Dewsbury, D. A. 1981. Editor's comments on papers 36 through 41. In *Mammalian sexual behavior,* ed. D. A. Dewsbury. Stroudsburg, Pa.: Hutchinson Ross.

Diamond, M. 1970. Intromission pattern and species vaginal code in relation to induction of pseudopregnancy. *Science* 169:995–997.

Dirsh, V. M. 1969. Acridoidea of the Galapagos Islands. *Bulletin of the British Museum of Natural History (Entomology).* 23:27–51.

Doane, C. C. 1968. Aspects of the mating behavior of the gypsy moth. *Annals of the Entomological Society of America* 61(3):768–773.

Dobzhansky, T. 1941. *Genetics and the origin of species.* New York: Columbia University Press.

Dolling, W. R. 1981. A rationalized classification of the burrower bugs (Cydnidae). *Systematic Entomology* 6:61–76.

Dominey, W. J. 1983. Sexual selection, additive genetic variance and the "phenotypic handicap." *Journal of Theoretical Biology* 101:495–502.

Doty, R. L. 1974. A cry for the liberation of the female rodent: courtship and copulation in Rodentia. *Psychological Bulletin* 81(3):159–172.

Dowling, H., and J. Savage. 1960. A guide to the snake hemipenis: a survey of basic structure and systematic characteristics. *Zoologica* 45(1):17–28.

Drew, G. 1911. Sexual activities of the squid, *Loligo pealei* (Les). *Journal of Morphology* 22:327–360.

Dritschilo, W., H. Cornell, D. Nafus, and B. OConnor. 1975. Insular biogeography: of mice and mites. *Science* 190:467–469.

Duellman, W. E. 1970. *The hylid frogs of Middle America.* Monographs of the Museum of Natural History University of Kansas, pp. 1–753.

Dugdale, J. S. 1971. Entomology of the Aucklands and other islands south of New Zealand: Lepidoptera, excluding non-crambine Pyralidae. *Pacific Insects Monographs* 27:55–172.

Duncan, C. J. 1975. Reproduction. In *Pulmonates,* ed. V. Fretter and J. Peake. New York: Academic Press.

Dustan, G. G. 1964. Mating behaviour of the oriental fruit moth, *Grapholita molesta* (Busck) (Lepidoptera: Olethreutidae). *Canadian Entomologist* 96:1087–1093.

Dybas, H., and L. Dybas. 1981. Coadaptation and taxonomic differentiation of sperm and spermathecae in featherwing beetles. *Evolution* 35(1):168–174.

Eberhard, W. G. 1974. The natural history and behaviour of the wasp *Trigonopsis cameronii* Kohl (Sphecidae). *Transactions of the Royal Entomological Society of London* 125(3):295–328.

——— 1979. The function of horns in *Podischnus agenor* (Dynastinae) and other beetles. In *Sexual selection and reproductive competition in insects,* ed. M. Blum and N. Blum. New York: Academic Press.

——— 1981. The natural history of *Doryphora* sp. (Coleoptera, Chrysomelidae) and the function of its sternal horn. *Annals of the Entomological Society of America* 74(5):445–448.

Eckstein, P., and S. Zuckerman. 1956. Morphology of the reproductive tract. In *Marshall's physiology of reproduction,* vol. 1, ed. A. S. Parkes. New York: Longman's Green and Co.

Economopoulos, A. P., and H. T. Gordon. 1972. Sperm replacement and depletion in the spermatheca of the S and CS strains of *Oncopeltus fasciatus. Entomologia Experimentalis et Applicata* 15:1–12.

Edgren, R. A. 1953. Copulatory adjustment in snakes and its evolutionary implications. *Copeia* 1953(3):162–164.

Edmunds, G., S. Jensen, and L. Berner. 1976. *The mayflies of North and Central America.* Minneapolis: University of Minnesota Press.

Edmunds, M. 1969. Opisthobranchiate Mollusca from Tanzania, I: Eolidacea (Eubranchidae and Aeolidiidae). *Proceedings of the Malacological Society of London* 38:451–469.

——— 1970. Opisthobranchiate Mollusca from Tanzania, II: Eolidacea (Cuthonidae, Piseinotecidae and Facelinidae). *Proceedings of the Malacological Society of London* 39:15–57.

Ehrlich, A. H., and P. R. Ehrlich. 1978. Reproductive strategies in the butterflies. I: Mating frequency, plugging, and egg number. *Journal of the Kansas Entomological Society* 51:666–697.

Ehrlich, P. R. 1965. The population biology of the butterfly, *Euphydryas editha,* II: the structure of the Jasper Ridge colony. *Evolution* 19:327–336.

El Said, A. 1976. Contribution to reproduction of *Amblyomma hebraeum* (Acari: Ixodidae): spermatogenesis and copulation. Ph.D. diss., University of Geneva.

Emerson, K. C., and R. D. Price. 1981. A host-parasite list of the Mallophaga on mammals. *Miscellaneous Publications of the Entomological Society of America* 12:1–72.

Englemann, F. 1970. *The physiology of insect reproduction.* New York: Pergamon Press.

Etman, A. A. M., and G. H. S. Hooper. 1979. Sperm precedence of the last mating in *Spodoptera litura*. *Annals of the Entomological Society of America* 72(1):119–120.

Evans, D.A., and R. W. Matthews. 1976. Comparative courtship behaviour in two species of the parasitic chalcid wasp *Melittobia* (Hymenoptera: Eulophidae). *Animal Behaviour* 24(1):46–51.

Evans, G. O., J. G. Sheals, and D. Macfarlane. 1961. *The terrestrial Acari of the British Isles*, vol. 1. London: British Museum.

Evans, H. E., and M. J. W. Eberhard. 1970. *The wasps*. Ann Arbor: University of Michigan Press.

Ewer, R. F. 1973. *The carnivores*. Ithaca, N.Y.: Cornell University Press.

Ewing, A. W. 1977. Communication in Diptera. In *How animals communicate*, ed. T. A. Sebeok. Bloomington: Indiana University Press.

Exline, H., and H. W. Levi. 1962. American spiders of the genus *Argyrodes* (Araneae Theridiidae). *Bulletin of the Museum of Comparative Zoology* 127(2):75–204.

Fain, A. 1967. Le genre *Dermatophagoides* Bogdanov 1864: son importance dans les allergies respiratoires cutanées chez l'homme (Psoroptidae: Sarcoptiformes). *Acarologia* 9(1):179–225.

——— 1981. Le genre *Listrophoroides* Hirst, 1923 (Acari, Astigmata, Atopomelidae) dans la region orientàle. *Bulletin de l'Institut Royal des Sciences Naturelles de Belgique (Bruxelles)* 53(6):1–123.

Fairchild, K. 1977. *The polychaete worms*. Natural History Museum of Los Angeles County, Allan Hancock Foundation Science Series 28:1–188.

Falconer, D. S. 1981. *Introduction to quantitative genetics*, 2d ed. London: Longman.

Fall, H. C. 1912. A review of the North American species of *Collops* (Col.). *Journal of the New York Entomological Society* 20:249–274.

Fatzinger, C. W., and W. C. Asher. 1971. Mating behavior and evidence for a sex pheromone of *Dioryctria abietella* (Lepidoptera: Pyralidae, Phycitinae). *Annals of the Entomological Society of America* 64:612–620.

Feldman-Muhsam, B. 1973. Copulation and spermatophore formation in soft and hard ticks. *Proceedings of the 3rd International Congress of Acarology, Prague, 1971:* 719–722.

——— 1979. Copulatory behavior and fecundity of male *Ornithodoros* ticks. *Recent Advances in Acarology* 2:159–166.

Feldman-Muhsam, B., S. Borut, and S. Saliternik-Givant. 1970. Salivary secretion of the male tick during copulation. *Journal of Insect Physiology* 16:1945–1949.

Fennah, R. G. 1945. The Cixiini of the lesser Antilles. *Proceedings of the Biological Society of Washington* 58:133–146.

——— 1946. On the formation of species and genera in the insect fauna of the lesser Antillean archipelago. *Proceedings of the Royal Entomological Society of London* (A) 21:73–79.

——— 1967. Fulgoroidea from the Galapagos archipelago. *Proceedings of the California Academy of Sciences*, 4th ser. 35:53–102.

Ferris, G. F. 1951. *The sucking lice.* Memoirs of the Pacific Coast Entomological Society 1:1–320.

Fisher, R. 1958. *The genetical theory of natural selection.* New York: Dover.

Foltz, D. W. 1981. Genetic evidence for long-term monogamy in a small rodent, *Peromyscus polionotus. American Naturalist* 117(5):665–675.

Fooden, A. 1970. Complementary specialization of male and female reproductive structures in the bear macaque, *Macaca arctoides. Nature* 214:939.

Forbes, W. T. M. 1941. Does he stridulate? (Lepidoptera: Eupterotidae). *Entomological News* 52:79–82.

Forrest, T. 1983. Calling songs and mate choice in male crickets. In *Orthopteran mating systems: sexual competition in a diverse group of insects,* ed. D. Gwynne and G. Morris, pp. 185–204. Boulder, Col.: Westview Press.

Forster, R. R. 1954. The New Zealand harvestmen (sub-order Laniatores). *Canterbury Museum Bulletin* 2:1–329.

Fowler, G. 1973. Some aspects of the reproductive biology of *Drosophila:* sperm transfer, sperm storage, and sperm utilization. *Advances in Genetics* 17:293–360.

Francke, O. F. 1979. Spermatophores of some North American scorpions (Arachnida, Scorpiones). *Journal of Arachnology* 7:19–32.

Franclemont, J. C. 1973. *The moths of America north of Mexico.* Fasc. 20, pt. 1: *Mimallonoidea and Bombycoidea.* London: E. W. Classey Ltd. and R. B. D. Publishers.

Fraser, F. C. 1943. The function and comparative anatomy of the oreillets in the Odonata. *Proceedings of the Royal Entomological Society of London* (A) 18:50–56.

Frechin, D. 1969. A notable intergeneric mating (Lycaenidae). *Journal of the Lepidopterist's Society* 23:115.

Freitag, R. 1974. Selection for non-genitalic mating structure in female tiger beetles of the genus *Cicindela* (Coleoptera: Cicindelidae). *Canadian Entomologist* 106(6):561–568.

Freitag, R., J. Olynyk, and B. Barnes. 1980. Mating behavior and genitalic counterparts in tiger beetles (Carabidae: Cicindelinae). *International Journal of Invertebrate Reproduction* 2:131–135.

Fretter, V., and A. Graham. 1964. Reproduction. In *Physiology of Mollusca,* ed. K. M. Wilbur and C. M. Yonge. New York: Academic Press.

Frey, G. 1973. Neue Macrodactylini (Col., Scarab., Melolonthinae). *Entomologische Arbeiten* 24:255–279.

Friedel, T., and C. Gillott. 1977. Contribution of male-produced proteins to vitellogenesis in *Melanoplus sanguinipes. Journal of Insect Physiology* 23:145–151.

Gagné, R.J., and R. D. Peterson. 1982. Physical changes in the genitalia of males of the screwworm *Cochliomyia hominovirax* (Diptera: Calliphoridae), caused by mating. *Annals of the Entomological Society of America* 75:574–578.

Garrison, R. W. 1982. *Archilestes neblina,* a new damselfly from Costa Rica, with comments on the variability of *A. latialatus* Donnelly (Odonata: Lestidae). *Occasional Papers of the Museum of Zoology* University of Michigan 702:1–12.

Garton, J. S. 1972. Courtship of the small-mouthed salamander, *Ambystoma texanum,* in southern Illinois. *Herpetologica* 28(1):41–45.

Gehring, R. D., and H. F. Madsen. 1963. Some aspects of the mating and oviposition behavior of the codling moth, *Carpocapsa pomonella. Journal of Economic Entomology* 56(2):140–143.

George, J. A., and M. G. Howard. 1968. Insemination without spermatophores in the oriental fruit moth, *Grapholita molesta* (Lepidoptera: Tortricidae). *Canadian Entomologist* 100:190–192.

Gerber, G. 1967. A possible mechanism for the regulation of the female reproductive cycle in *Tenebrio molitor* (Coleoptera: Tenebrionidae). *Canadian Entomologist* 99:1298–1303.

Gering, R. L. 1953. Structure and function in the genitalia in some American agelenid spiders. *Smithsonian Miscellaneous Collections* 121(4):1–84.

Gertsch, W. J. 1949. *American spiders.* New York: van Nostrand.

——— 1964. The spider genus *Zygiella* in North America (Araneae, Argiopidae). *American Museum Novitates* 2188:1–21.

Gertsch, W. J., and F. Ennik. 1983. The spider genus *Loxosceles* in North America, Central America, and the West Indies (Araneae, Loxoscelidae). *Bulletin of the American Museum of Natural History* 175:264–360.

Ghirardelli, E. 1968. Some aspects of the biology of the chaetognaths. *Advances in Marine Biology* 6:271–375.

Gibbons, L. M. 1978. Revision of the genus *Paracooperia* Travassos, 1935 (Nematoda: Trichostrongylidae). *Journal of Helminthology* 52:231–249.

Gilbert, L. 1976. Postmating female odor in *Heliconius* butterflies: a male-contributed antiaphrodisiac? *Science* 193:419–420.

Gladney, W. J., and R. O. Drummand. 1971. Spermatophore transfer and fertilization of lone star ticks off the host. *Annals of the Entomological Society of America* 64(2):378–381.

Glasgow, H. 1933. The host relations of our cherry fruit flies. *Journal of Economic Entomology* 26:431–438.

Goldschmidt, R. 1940. *The material basis of evolution.* New Haven: Yale University Press.

Gordon, H. T., and W. Loher. 1968. Egg production and male activation in new laboratory strains of the large milkweed bug, *Oncopeltus fasciatus. Annals of the Entomological Society of America* 61(6):1573–1578.

Gordon, M., and D. E. Rosen. 1951. Genetics of species differences in the morphology of the male genitalia of xiphophorin fishes. *Bulletin of the American Museum of Natural History* 95:409–464.

Gordon, R. D. 1980. The tribe Azyini (Coleoptera: Coccinellidae): historical review and taxonomic revision. *Transactions of the American Entomological Society* 106:149–203.

Graf, J. F. 1978. Copulation, nutrition et ponte chez *Ixodes ricinus* L. (Ixodoidea: Ixodidae), II. *Mitteilungen der Schweizerischen Entomologischen Gesellschaft* 51:241–253.

Graham, H. M., P. A. Glick, M. T. Ouye, and D. F. Martin. 1965. Mating frequency of female pink bollworm collected from light traps. *Annals of the Entomological Society of America* 58:595–596.

Grasshoff, M. 1964. Die Kreuzspinne *Araneus pallidus*—ihr Netzbau und ihre Paarungsbiologie. *Natur und Museum* 94(8):305–314.

——— 1968. Morphologische Kriterien als Ausdruck von Artgrenzen bei Radnetzspinnen der Subfamilie Araneinae. *Abteilungen Senckenbergischen Naturforschenden Gesellschaft* 516:1–100.

——— 1973. Konstruktions-und Funktionanalyse an Kopulationsorganen einiger Radnetzspinnen. *Senckenbergische Naturforschende Gesellschaft in Frankfurt am Main. Bericht* 24:129–151.

——— 1974. Transformierungsreihen in der Stammesgeschichtemechanischer Wandel an Kopulationsorganen von Radnetzspinnen. *Natur und Museum* 104:321–330.

Green, J. W. 1957. Revision of the nearctic species of *Pyractomena* (Coleoptera: Lampyridae). *Wasmann Journal of Biology* 15(2):237–284.

Greenberg, B. 1943. Social behavior of the western banded gecko, *Coleonyx variegatus* Baird. *Physiological Zoology* 16(1):110–122.

Greenwood, P. H. 1975. *A history of fishes*. New York: Wiley and Sons.

Gressitt, J. L. 1962. Insects of Macquarie Island: introduction. *Pacific Insects* 4:905–916.

Gressitt, J. L., and K. A. J. Wise. 1971. Entomology of the Aucklands and islands south of New Zealand: introduction. *Pacific Insects Monographs* 27:1–46.

Griffiths, D. A., and J. Boczek. 1977. Spermatophores of some acaroid mites (Astigmata: Acarina). *International Journal of Insect Morphology and Embryology* 6(5–6):231–238.

Griswold, T. 1983. Revision of *Proteriades* subgenus *Acrosmia* Michener (Hymenoptera; Megachilidae). *Annals of the Entomological Society of America* 76:707–714.

Gromko, M., and D. Pyle. 1978. Sperm competition, male fitness, and repeated mating by female *Drosophila melanogaster*. *Evolution* 32(3):588–593.

Gross, M., and R. Shine. 1981. Parental care and mode of fertilization in ectothermic vertebrates. *Evolution* 35:775–793.

Gwadz, R., G. Craig, and W. Hickey. 1971. Female sexual behavior as the mechanism rendering *Aedes aegypti* refractory to insemination. *Biological Bulletin* (Woods Hole) 140:201–214.

Gwynne, D. T. 1982. Mate selection by female katydids (Orthoptera: Tettigoniidae, *Conocephalus nigropleurum*). *Animal Behaviour* 30:734–738.

——— 1983. Male nutritional investment and the evolution of sexual differences in Tettigoniidae and other Orthoptera. In *Orthopteran mating systems: sexual selection in a diverse group of insects,* ed. D. Gwynne and G. Morris, pp. 337–366. Boulder, Colo.: Westview Press.

—— Manuscript. Male nutrient investment, population density and sexual selection in mormon crickets *(Anabrus simplex,* Orthoptera: Tettigoniidae).

Haacker, U. 1969. An attractive secretion in the mating behaviour of a millipede. *Zeitschrift für Tierpsychologie* 26:988–990.

—— 1971. Die Funktion eines Drusenkomplexes im Balzverhalten von *Chordeuma* (Diplopoda). *Forma et Functio* 4:162–170.

—— 1974. Patterns of communication in courtship and mating behaviour of millipedes (Diplopoda). *Symposium of the Zoological Society of London* 32:317–328.

Haacker, U., and S. Fachs. 1970. Das Paarungsverhalten von *Cylindroiulus punctatus* Leach. *Zeitschrift für Tierpsychologie* 27:641–648.

Hafez, E. S. E. 1973. The comparative anatomy of the mammalian cervix. In *The biology of the cervix,* ed. R. J. Blandau and K. Moghissi. Chicago: University of Chicago Press.

Halffter, G., and E. G. Matthews. 1966. The natural history of dung beetles of the subfamily Scarabaeinae (Coleoptera, Scarabaeidae). *Folia Entomologica Mexicana* 14–16:1–287.

Halliday, T. R. 1977. The courtship of European newts: an evolutionary perspective. In *The reproductive biology of amphibians,* ed. D. H. Taylor and S. I. Guttman. New York: Plenum.

—— 1980. *Sexual strategy.* Chicago: University of Chicago Press.

Halstead, D. G. H. 1969. A new species of *Tribolium* from North America previously confused with *Tribolium madens* (Charp.) (Coleoptera: Tenebrionidae). *Journal of Stored Product Research* 4:295–304.

Hamilton, W. D. 1964. The genetical theory of social behavior, I, II. *Journal of Theoretical Biology* 7:1–52.

—— 1967. Extraordinary sex ratios. *Science* 156:477–488.

—— 1979. Wingless and fighting males in fig wasps and other insects. In *Sexual selection and reproductive competition in insects,* ed. M. and N. Blum. New York: Academic Press.

Hammond, P. M. 1981. The origin and development of reproductive barriers. In *The evolving biosphere.* Cambridge: Cambridge University Press.

Hand, C., and J. Steinberg. 1955. On the occurrence of the nudibranch *Alderia modesta* (Loven, 1844) on the central Californian coast. *Nautilus* 69(1):22–28.

Hanken, J., and P. W. Sherman. 1981. Multiple paternity in Belding's ground squirrel litters. *Science* 212:351–353.

Happ, G. 1969. Multiple sex pheromones of the mealworm beetle, *Tenebrio molitor* L. *Nature* 222:180–181.

Hardy, D. E. 1965. *Insects of Hawaii, vol. 12. Diptera: Cyclorrhapha, II. Series Schizophora, Section Acalypterae, I. Family Drosophilidae.* Honolulu: University of Hawaii Press.

Hartenstein, R. 1962. Life history studies of *Pergamasus crassipes* and *Amblygamasus septentrionalis* (Acarina: Parasitidae). *Annals of the Entomological Society of America* 55:196–202.

Hartman, O. 1965. *Deep-water benthic polychaetous annelids off New England to Bermuda and other North Atlantic areas.* Allan Hancock Foundation Occasional Papers 28:1–378.

—— 1967. *Polychaetous annelids collected by the USNS Eltanin and Staten Island cruises, chiefly from antarctic seas.* Allan Hancock Monographs in Marine Biology 2:1–387.

Haven, N. 1977. Cephalopoda: Nautiloidea. In *Reproduction of marine invertebrates,* vol. 4, ed. A. Giese and J. Pearse. New York: Academic Press.

Hayes, A. H. 1975. The larger moths of the Galapagos Islands (Geometroidea: Sphingoidea and Noctuoidea). *Proceedings of the California Academy of Sciences,* 4th ser. 40:145–208.

Hedgpeth, J. W. 1963. Pycnogonida of the North American Arctic. *Journal of the Fisheries Research Board of Canada* 20:1315–1348.

Helle, W. 1967. Fertilization in the two-spotted spidermite (*Tetranychus urticae:* Acari). *Entomologica Experimentalis et Applicata* 10:103–110.

Hellenthal, R. A., and R. D. Price. 1976. Louse-host associations of *Geomydoecus* (Mallophaga: Trichodectidae) with yellow-faced pocket gopher, *Pappageomys castanops* (Rodentia: Geomyidae). *Journal of Medical Entomology* 13:331–336.

Hendelberg, J. 1974. Spermiogenesis, sperm morphology, and biology of fertilization in the Turbellaria. In *Biology of the Turbellaria,* ed. N. Riser and P. Morse. New York: McGraw-Hill.

Henley, C. 1974. Platyhelminthes. In *Reproduction of marine invertebrates,* vol. 1, ed. A. Giese and J. Pearse. New York: Academic Press.

Hennig, W. 1949. Sepsidae. In *Die Fliegen der Paleartischen Region,* no. 39a, ed. Lindner, pp. 1–91.

Hershkovitz, P. 1979. *Living new world monkeys (Platyrrhini) with an introduction to primates,* vol. 1. Chicago: University of Chicago Press.

Hewer, H. R. 1934. Studies in *Zygaena* (Lepidoptera), II: the mechanism of copulation and the passage of the sperm in the female. *Proceedings of the Zoological Society of London* 104:513–527.

Heyer, W. R. 1970. Studies on the genus *Leptodactylus* (Amphibia: Leptodactylidae), II: diagnosis and distribution of the *Leptodactylus* of Costa Rica. *Revista de Biología Tropical* 16(2):171–205.

Higgins, L. G. 1960. A revision of the melitaeine genus *Chlosyne* and allied species (Lepidoptera: Nymphalinae). *Transactions of the Royal Entomological Society of London* 112:381–467.

—— 1981. A revision of *Phycioides* Hubner and related genera, with a review of the classification of the Melitaeinae (Lepidoptera: Nymphalidae). *Bulletin of the British Museum of Natural History (Entomology)* 43:77–243.

Hippa, H., and I. Oksala. 1983. Epigynal variation in *Enoplognatha latimana* Hippa and Oksala (Araneae, Theridiidae) in Europe. *Bulletin of the British Arachnological Society* 6:99–102.

Hirai, K., H. H. Shorey, and L. K. Gaston. 1978. Competition among courting male moths: male-to-male inhibitory pheromone. *Science* 202:644–645.

Hooper, E., and G. Musser. 1964. The glans penis in neotropical cricetines (Muridae) with comments on classification of muroid rodents. *Miscellaneous Publications of the Museum of Zoology* University of Michigan 123:1–57.

Hope, W. D. 1974. Nematoda. In *Reproduction of Marine Invertebrates,* vol. 1, ed. A. Giese and J. Pearse. New York: Academic Press.

Hopkins, G. H. E. 1949. The host-associations of the lice of mammals. *Proceedings of the Zoological Society of London* 119:387–604.

Howden, H. G. 1979. A revision of the Australian genus *Blackburnium* Boucomont (Coleoptera: Scarabeidae: Geotrupinae). *Australian Journal of Zoology* (Supp. Ser.) 72:1–88.

Hoyle, W. E. 1907. Presidential address. *Report of the British Association for the Advancement of Science* 77:520–539.

Hummon, W. D. 1974. *Gastrotricha.* In *Reproduction of marine invertebrates,* vol. 4, ed. A. Giese and J. Pearse, pp. 485–506. New York: Academic Press.

Hungerford, H. B. 1954. The genus *Rheumatobates* Bergroth (Hemiptera-Gerridae). *University of Kansas Science Bulletin* 36(1):529–588.

Hunter, R. H. F. 1975. Transport, migration and survival of spermatozoa in the female genital tract: species with intra-uterine deposition of semen. In *The biology of spermatozoa,* ed. E. S. E. Hafez and C. G. Thibault. New York: S. Karger.

Hyman, L. H. 1940. *The invertebrates: Protozoa through Ctenophora.* New York: McGraw-Hill.

—— 1951a. *The invertebrates: Platyhelminthes and Rhynchocoela.* New York: McGraw-Hill.

—— 1951b. *The invertebrates: Acanthocephala, Aschelminthes, and Entoprocta.* New York: McGraw-Hill.

—— 1955. *The invertebrates: Echinodermata.* New York: McGraw-Hill.

—— 1959. *The invertebrates: the smaller coelomate groups.* New York: McGraw-Hill.

Imms, A. D. 1957. *A general textbook of entomology,* rev. O. W. Richards and R. G. Davies. New York: Barnes and Noble.

Inger, R. F., and B. Greenberg. 1956. Morphology and seasonal development of sex characters in two African toads. *Journal of Morphology* 99:549–574.

Inglis, W. G. 1961. The oxyurid parasites (Nematoda) of primates. *Proceedings of the Zoological Society of London* 136:103–122.

Ishiyama, R. 1967. *Fauna Japonica: Rajidae (Pisces).* Tokyo: Biogeographical Society of Japan.

Jackson, R. R. 1980. The mating strategy of *Phidippus johnsoni* (Araneae, Salticidae), II: sperm competition and the function of copulation. *Journal of Arachnology* 8:217–240.

—— 1981. Relationship between reproductive security and intersexual selection in a jumping spider *Phidippus johnsoni* (Araneae: Salticidae). *Evolution* 35:601–604.

Jeannel, R. 1941. L'isolement, facteur de l'évolution. *Revue Francaise d'Entomologie* 8(3):101–110.

——— 1955. *L'Édéage*. Paris: Éditions du Muséum.

Johnson, C. 1972. Tandem linkage, sperm translocation, and copulation in the dragonfly, *Hagenius brevistylus* (Odonata: Gomphidae). *American Midland Naturalist* 88(1):131–149.

Johnson, C. D. 1983. Ecosystematics of *Acanthoscelides* (Coleoptera: Bruchidae) of southern Mexico and Central America. *Miscellaneous Publications of the Entomological Society of America* 56:1–248.

Johnson, P. 1972a. *Neohaematopinus appressus,* a new species of sucking louse from an Asian tree squirrel (Anoplura). *Pacific Insects* 14:389–392.

——— 1972b. Sucking lice of Venezuelan rodents, with remarks on related species (Anoplura). *Brigham Young University Bulletin* 17(5):1–62.

Jordan, E. O. 1891. The spermatophores of *Diemyctylus*. *Journal of Morphology* 5:263.

Joyeux, C., and J.-G. Baer. 1961. Classe des Cestodaires. In *Traité de zoologie*, vol. 4, fasc. 1, ed. P. P. Grassé. Paris: Masson.

Kaestner, A. 1968. *Invertebrate zoology*, trans. and adapt. H. Levi and L. Levi. New York: Wiley and Sons.

Kaneshiro, K. K. 1983. Sexual selection and direction of evolution in the biosystematics of Hawaiian Drosophilidae. *Annual Review of Entomology* 28:161–178.

Kaston, B. J. 1948. *The spiders of Connecticut*. Bulletin of the Connecticut State Geological and Natural History Survey 70:1–874.

Kemp, D. H., B. F. Stone, and K. C. Binnington. 1982. Tick attachment and feeding: role of the mouthparts, feeding apparatus, salivary gland secretions and the host response. In *Physiology of Ticks,* ed. F. D. Obenchain and R. Galun. New York: Pergamon Press.

Kennedy, C. H. 1919. A study of the phylogeny of the Zygoptera from evidence given by the genitalia. Ph.D. diss., Cornell University.

——— 1920. The phylogeny of the zygopterous dragonflies as based on the evidence of the penes. *Ohio Journal of Science* 22(1):19–29.

Kenney, A. M., D. L. Lanier, and D. A. Dewsbury. 1977. Effects of vaginal-cervical stimulation in seven species of muroid rodents. *Journal of Reproduction and Fertility* 49:305–309.

Kerkis, J. 1931. Vergleichende Studien über die Variabilität der Merkmale des Geschlechtsaparats und der äusseren Merkmale bei *Eurygaster integriceps* Put. *Zoologischer Anzeiger* 14:129–143.

Kernbach, K. 1962. Die Schwarmer einiger Galapagos-Inseln. *Opuscula Zoologica* 63:1–19.

Kerr, W., R. Zucchi, J. Nakadaira, and J. Butolo. 1962. Reproduction in the social bees (Hymenoptera: Apidae). *Journal of the New York Entomological Society* 70:265–276.

Key, K. H. L. 1981. Species, parapatry, and the morabine grasshoppers. *Systematic Zoology* 30:425–458.

Kim, K. C. 1965. A review of the *Hoplopleura hesperomydis* complex (Anoplura, Hoplopleuridae). *Journal of Parasitology* 51:871–887.

—— 1966. The species of *Enderleinellus* (Anoplura, Hoplopleuridae) parasitic on the Sciurini and Tamiasciurini. *Journal of Parasitology* 52:988–1023.

Kim, K. C., and C. F. Weisser. 1974. Taxonomy of *Solenopotes* Enderlein, 1904, with redescription of *Linognathus panamensis* Ewing (Linognathidae: Anoplura). *Parasitology* 69:107–135.

Kimura, M. 1979. The neutral theory of molecular evolution. *Scientific American* (Nov.):94–104.

King, P. E. 1973. *Pycnogonids*. London: Hutchinson.

Kistner, D. H. 1966. A revision of the African species of the Aleocharine tribe Dorylomimini (Coleoptera: Staphylinidae). II: The genera *Dorylomimus, Dorylonannus, Dorylogaster, Dorylobactrus,* and *Mimanomma,* with notes on their behavior. *Annals of the Entomological Society of America* 59:320–340.

Klauber, L. 1972. *Rattlesnakes,* vol. 1. Los Angeles: University of California Press.

Kleiman, D. 1977. Monogamy in mammals. *Quarterly Review of Biology* 52:39–69.

Klingel, H. 1956. Indirekte Spermatophorenubertrugung bei Chilopoden (Hundertfusser) beobachtet bei der "Spinnenassel" *Scutigera coleoptrata* Latzel. *Naturwissenschaft* 43:311.

—— 1957. Indirekte Spermatophorenubertragung beim scolopender (*Scolopendra cingulata* Latrielle; Chilopoda, Hundertfusser). *Naturwissenschaft* 44:338.

—— 1959. Indirekte Spermatophorenubertragung bei Geophiliden (Hundertfusser, Chilopoda). *Naturwissenschaft* 46:632–633.

—— 1962. Das paarungsverhalten des malaischen Hohlentrausendfusses *Thereuopoda decipiens cavernicola* Verhoeff (Scutigeromorpha, Chilopoda). *Zoologische Anzeiger* 169:458–460.

Klots, A. B. 1970. Lepidoptera. In *Taxonomist's glossary of genitalia in insects,* ed. S. L. Tuxen. Darien, Conn.: S-H Service Agency.

Kluge, A. In prep. Cloacal bones and sacs as evidence of gekkonid lizard relationships.

Koeniger, G. 1981. In welchem Abschnitt des Paarungsverhaltens der Bienenkonigin findet die Induktion der Eiablage statt? *Apidologie* 12(4):329–343.

—— 1983. Die Entfernung des Begattungszeichens bei der Mehrfachpaarung der Bienenkonigin. *Allgemeine Deutsche Imker* (Aug.):244–245.

Kopp, D. D., and T. R. Yonke. 1973. The treehoppers of Missouri, II: subfamily Smiliinae; tribes Acutalini, Ceresini, and Polyglyptini. *Journal of the Kansas Entomological Society* 46:233–276.

Kormondy, E. J. 1959. The systematics of *Tetragoneuria*, based on ecological, life history, and morphological evidence (Odonata: Corduliidae). *Miscellaneous Publications of the Museum of Zoology* University of Michigan 107:1–79.

Kraus, O. 1968. Isolationsmechanismen und Genitalstrukturen bei wirbellosen Tieren. *Zoologische Anzeiger* 181:22–38.

Krieger, F., and E. Krieger-Loibl. 1958. Beitrage zum Verhalten von *Ischnura elegans* und *Ischnura pumilio* (Odonata). *Zeitschrift für Tierpsychologie* 15:82–93.

Kritsky, D. C. and V. E. Thatcher. 1974. Monogenic trematodes (Monophisthocotylea: Dactylogyridae) from freshwater fishes of Colombia. *Journal of Helminthology* 48:59–66.

Krutzch, P. H., and T. A. Vaughn. 1955. Additional data on the bacula of North American bats. *Journal of Mammology* 36:96–100.

Krzywinski, A., and Z. Jaczewski. 1978. Observations on the artificial breeding of red deer. *Symposium of the Zoological Society of London* 43:271–287.

Kullman, E. 1964. Neue Ergebnisse uber den Netzbau und das Sexualverhalten einiger Spinnenarten. *Zeitschrift für zoologische Systematik und Evolutionsforschung* 2:41–122.

Kunze, L. 1959. Die functionsanatomischen Grundlagen der Kopulation Zwerzikaden, untersucht an *Euscelis plebejus* (Fall.) und einigen Typhlocybinen. *Deutsche Entomologische Zeitschrift* 6(4):322–387.

Labine, P. A. 1966. The population biology of the butterfly, *Euphydryas editha*, IV: sperm precedence—a preliminary report. *Evolution* 20:580–586.

Lambert, R. D., and C. Tremblay. 1978. Effet du coït sur la montée des spermatozoïdes et la fécondation chez le lapin. *Revue Canadienne de Biologie* 37(1):1–4.

Lande, R. 1981. Models of speciation by sexual selection on polygenic traits. *Proceedings of the National Academy of Sciences* (U.S.A.) 78:3721–3725.

Lasserre, P. 1975. Clitellata. In *Reproduction of marine invertebrates*, vol. 3, ed. A. Giese and J. Pearse. New York: Academic Press.

Lavigne, R. 1970a. Courtship and predatory behavior of *Cyrtopogon auratus* and *C. glarealis* (Diptera: Asilidae). *Journal of the Kansas Entomological Society* 43(2):163–171.

——— 1970b. Courtship and predation behavior of *Heteropogon maculinervis* (Diptera: Asilidae). *Journal of the Kansas Entomological Society* 43(3):270–273.

——— 1972. Ethology of *Ablautus rufotibialis* on the Pawnee grasslands IBP site. *Journal of the Kansas Entomological Society* 45(3):271–274.

Lea, A. 1968. Mating without insemination in virgin *Aedes aegypti*. *Journal of Insect Physiology* 14:305–308.

Leahy, M. G. 1967. Non-specificity of the male factor enhancing egg-laying in Diptera. *Journal of Insect Physiology* 13:1283–1292.

Lee, D. C. 1970. The Rhodacaridae (Acari: Mesostigmata): classification, external morphology and distribution of genera. *Records of the South Australian Museum* 16(3):1–219.

——— 1974. Rhodacaridae (Acari: Mesostigmata) from near Adelaide, Australia, III: behaviour and development. *Acarologia* 16:21–44.

Lefevre, G., and U. Jonsson. 1962. Sperm transfer, storage, displacement, and utilization in *Drosophila melanogaster. Genetics* 47:1719–1736.

Leigh-Sharpe, W. H. 1920. The comparative morphology of the secondary sexual characters of elasmobranch fishes, I: the claspers, clasper siphons, and clasper glands. *Journal of Morphology* 34(2):245–265.

——— 1922. The comparative morphology of the secondary sexual characters of elasmobranch fishes. *Journal of Morphology* 36(2):191–197, 198–219, 221–243.

Leopold, R. A., and M. E. Degrugillier. 1973. Sperm penetration of housefly eggs: evidence for involvement of a female accessory secretion. *Science* 181:555–557.

Leslie, J. F., and H. Dingle. 1983. Interspecific hybridization and genetic divergence in milkweed bugs (*Oncopeltus:* Hemiptera: Lygaeidae). *Evolution* 37:583–591.

Levi, H. W. 1959. The spider genera *Achaearanea, Theridion* and *Sphyrotinus* from Mexico, Central America and the West Indies (Araneae, Theridiidae). *Bulletin of the Museum of Comparative Zoology* 121(3):57–163.

——— 1968. The spider genera *Gea* and *Argiope* in America (Araneae: Araneidae). *Bulletin of the Museum of Comparative Zoology* 136(9):319–352.

——— 1971. The *diadematus* group of the orb-weaver genus *Araneus* north of Mexico (Araneae: Araneidae). *Bulletin of the Museum of Comparative Zoology* 141(4):131–179.

——— 1974. The orb-weaver genus *Zygiella* (Araneae: Araneidae). *Bulletin of the Museum of Comparative Zoology* 146(5):267–290.

——— 1975. Mating behavior and presence of embolus cap in male Araneidae. *Proceedings of the Sixth International Arachnological Congress,* 49–50.

——— 1977. The American orb-weaver genera *Cyclosa, Metazygia* and *Eustala* north of Mexico (Araneae, Araneidae). *Bulletin of the Museum of Comparative Zoology* 148(3):61–127.

——— 1978. Orb webs and phylogeny of orb-weavers. *Symposium of the Zoological Society of London* 42:1–15.

——— 1981. The American orb-weaver genera *Dolichognatha* and *Tetragnatha* north of Mexico (Araneae: Araneidae, Tetragnathinae). *Bulletin of the Museum of Comparative Zoology* 149(5):271–318.

Levine, L., M. Asmussen, O. Olvera, J. R. Powell, M. E. de la Rosa, V. M. Salceda, M. I. Gaso, J. Guzman, and W. W. Anderson. 1980. Population genetics of Mexican *Drosophila,* V: a high rate of multiple insemination in a natural population of *Drosophila pseudoobscura. American Naturalist* 116:493–503.

Lindroth, C. H., and E. Palmen. 1970. Coleoptera. In *Taxonomist's glossary of insect genitalia*, ed. S. Tuxen. Darien, Conn.: S-H Service Agency.

Lindsey, A. 1939. Variations of insect genitalia. *Annals of the Entomological Society of America* 32:173–176.

Linley, J. R. 1975. Sperm supply and its utilization in doubly inseminated flies, *Culicoides melleus. Journal of Insect Physiology* 21:1785–1788.

Lipovsky, L. J., G. W. Byers, and E. H. Kardos. 1957. Spermatophores—the mode of insemination of chiggers (Acarina: Trombiculidae). *Journal of Parasitology* 43(1):256–262.

Lloyd, J. E. 1979. Mating behavior and natural selection. *Florida Entomologist* 62(1):17–23.

———— 1983. Bioluminescence and communication in insects. *Annual Review of Entomology* 28:131–160.

Löbl, I. 1977. *Baeocera galapagoensis* nov. spe., a new scaphidiid beetle from the Galapagos Islands (Coleoptera, Scaphidiidae). *Studies on Neotropical Fauna and Environment.* 12:249–252.

Loibl, E. 1958. Zur Ethologie und Biologie der deutschen Lestiden (Odonata). *Zeitschrift für Tierpsychologie* 15:54–81.

Long, C. A., and T. Frank. 1968. Morphometric variation and function in the baculum, with comments on correlation of parts. *Journal of Mammology* 49:32–45.

Lopes, H. de S. 1978. Sarcophagidae (Diptera) of Galapagos Islands. *Revista Brasiliera de Biología* 38:595–611.

Lopez, A., and M. Emerit. 1979. Donnes complementaires sur la glande clypeale des *Argyrodes* (Araneae, Theridiidae): utilisation du microscope electronique à balayage. *Revue Arachnologique* 2:143–153.

Lorkovic, Z. 1952. L'accouplement artificiel chez les Lépidoptères et son application dans les recherches sur la fonction de l'appareil génital des insectes. *Physiological and Comparative Oecology* 3:313–319.

Lourenço, W. R. 1980. A proposito de duas novas especies de *Opisthacanthus* para a regiâo neotropical: *Opisthacanthus valerioi* da "Isla del Coco," Costa Rica e *Opisthacanthus heurtaultae* da Guiana Francesa (Scorpiones, Scorpionidae). *Revista Nordeste de Biologia* 3(2):179–194.

Maa, T. C. 1971. Review of the Streblidae (Diptera) parasitic on megachiropteran bats. *Pacific Insects Monographs* 18:213–243.

Makielski, S. K. 1972. Polymorphism in *Papilio glaucus* L. (Papilionidae): maintenance of the female ancestral form. *Journal of the Lepidopterist's Society* 20:109–111.

Mann, K. H. 1962. *Leeches (Hirudinea): their structure, physiology, ecology, and embryology.* New York: Pergamon Press.

Mann, T., and L. C. Prosser. 1963. Uterine response to 5-hydroxytryptamine in the clasper-siphon secretion of the spiny dogfish *Squalus acanthias. Biological Bulletin* (Woods Hole) 125:384–385.

Manning, A. 1966. Sexual behavior. *Symposium of the Royal Entomological Society of London* 3:59–68.

Markow, T. A. 1981. Courtship behavior and control of reproductive isolation between *Drosophila mojavensis* and *Drosophila arizonensis*. Evolution 35:1022–1026.

Marks, R. J. 1976. Mating behavior and fecundity of the red bollworm *Diparopsis castanea* Hmps. (Lepidoptera, Noctuidae). *Bulletin of Entomological Research* 66:145–158.

Martin, C. H. 1968. The new family Leptogastridae (the grass flies) compared with the Asilidae (robber flies) (Diptera). *Journal of the Kansas Entomological Society* 41(1):70–100.

Martin, C. O., and D. J. Schmidly. 1982. Taxonomic review of the pallid bat *Antrozous pallidus* (LeConte). *Special Publications The Museum Texas Tech University* 18:1–48.

Martyniuk, J., and J. Jaenike. 1982. Multiple mating and sperm usage patterns in natural populations of *Prolinyphia marginata* (Araneae: Linyphiidae). *Annals of the Entomological Society of America* 75:516–518.

Mason, J. C. 1970. Copulatory behavior of the crayfish, *Pacifastacus trowbridgii* (Stimpson). *Canadian Journal of Zoology* 48:969–976.

Massoud, Z. and J.-M. Betsch. 1972. Étude sur les insectes collemboles, II: les caractères sexuels secondaires des antennes des Symphypléones. *Revue d'Ecologie et de Biologie du Sol* 9:55–97.

Masters, W. H., and V. E. Johnson. 1966. *The human sexual response*. New York: Little, Brown.

Mather, S. N., and E. J. LeRoux. 1970. The reproductive organs of the velvet mite, *Allothrombium lerouxi* (Thrombidiformes: Thrombidiidae). *Canadian Entomologist* 102:144–157.

Matsuda, R. 1960. Morphology, evolution and a classification of the Gerridae (Hemiptera-Heteroptera). *University of Kansas Science Bulletin* 41:25–632.

Matthews, R. W., A. Hook, and J. W. Krispin. 1979. Nesting behavior of *Crabro argusinus* and *C. hilaris* (Hymenoptera, Sphecidae). *Psyche* 86(2–3):149–166.

Mawson, P. M. 1977. The genus *Microtetrameres* Travassos (Nematoda: Spirurida) in Australian birds. *Records of the South Australian Museum* 17:239–259.

Mayer, H. 1957. Zur Biologie und Ethologie einheimischer Collembolen. *Zoologische Jahrbüchen Abteilung für Systematik* 85(6):501–570.

Mayr, E. 1963. *Animal species and evolution*. Cambridge, Mass.: Harvard University Press.

McGill, T. E. 1970. Induction of luteal activity in female house mice. *Hormones and Behavior* 1:211–222.

——— 1977. Reproductive isolation, behavioral genetics, and functions of sexual behavior in rodents. In *Reprodutive behavior and evolution*, ed. J. S. Rosenblatt and B. R. Komisaruk. New York: Plenum Press.

McLain, D. K. 1980. Female choice and the adaptive significance of prolonged copulation in *Nezara viridula* (Hemiptera: Pentatomidae). *Psyche* 87(3–4):325–336.

Meglitsch, P. A. 1967. *Invertebrate zoology.* Cambridge: Oxford University Press.

Meijer, J. 1977. A glandular secretion in the ocular area of certain erigonine spiders (Araneae, Linyphiidae). *Bulletin of the British Arachnological Society* 3:251–252.

Menke, A. S. 1960. A taxonomic study of the genus *Adebus* Stål (Hemiptera, Belostomatidae). *University of California Publications in Entomology* 16(8):393–440.

Merritt, R., and B. Wu. 1975. On the quantification of promiscuity in *Peromyscus maniculatus. Evolution* 29:575–578.

Metcalf, R. A., and G. S. Whitt. 1977. Intra-nest relatedness in the social wasp *Polistes metricus. Behavioral Ecology and Sociobiology* 2:339–351.

Michener, C. 1974. *The social behavior of the bees.* Cambridge, Mass.: Harvard University Press.

Milkman, R., and R. Zeitler. 1974. Concurrent multiple paternity in natural and laboratory populations of *Drosophila melanogaster. Genetics* 78:1191–1193.

Milledge, A. F. 1980. The erigonine spiders of North America, II: the genus *Spirembolus* Chamberlin (Araneae: Linyphiidae). *Journal of Arachnology* 8:109–158.

——— 1981a. The erigonine spiders of North America, III: the genus *Scotinotylus* Simon (Araneae: Linyphiidae). *Journal of Arachnology* 9:167–213.

——— 1981b. The erigonine spiders of North America, IV: the genus *Disembolus* Chamberlin and Ivie (Araneae: Linyphiidae). *Journal of Arachnology* 9:259–284.

Miller, C. 1970. The nearctic species of *Pnigalio* and *Sympiesis* (Hymenoptera: Eulophidae). *Memoirs of the Entomological Society of Canada* 68:1–121.

Mitchell, R. 1957. The mating behavior of pionid water-mites. *American Midland Naturalist* 58(2):360–366.

——— 1958. Sperm transfer in the water mite *Hydryphantes ruber* Geer. *American Midland Naturalist* 60(1):156–157.

Moghissi, K. S. 1971. Sperm migration through cervical mucus. In *Pathways to conception,* ed. A. Sherman. Springfield, Ill.: Thomas.

Mohr, C. E. 1931. Observations of the early breeding habits of *Ambystoma jeffersonianum* in central Pennsylvania. *Copeia* 1931(3):102–104.

Morris, G. K., and J. H. Fullard. 1983. Random noise and congeneric discrimination in *Conocephalus* (Orthoptera: Tettigoniidae). In *Orthopteran mating systems,* ed. D. T. Gwynne and G. K. Morris. Boulder Colo.: Westview Press.

Morse, J. C. 1972. The genus *Nyctiophylax* in North America. *Journal of the Kansas Entomological Society* 45(2):172–181.

Moss, W. W. 1960. Description and mating behaviour of *Allothrombium lerouxi,* new species (Acarina: Thrombiidae), a predator of small arthropods in Quebec apple orchards. *Annals of the Entomological Society of America* 92:898–905.

Moynihan, M., and A. F. Rodaniche. 1982. The behavior and natural history of the Caribbean reef squid *Sepioteuthis sepioidea*. *Advances in Ethology* 25:1–150.

Mroczkowski, M. 1966. Contribution to the knowledge of Silphidae and Dermestidae of Korea (Coleoptera). *Annales Zoologici* (Warsaw) 23(16):1–10.

Murray, J. 1964. Multiple mating and effective population size in *Cepea nemoralis*. *Evolution* 18:283–291.

Myers, C. W. 1974. The systematics of *Rhadinaea* (Colubridae), a genus of new world snakes. *Bulletin of the American Museum of Natural History* 153:1–262.

Nault, L. R. 1984. Corn leafhopper: the making of an insect pest. *Ohio Report* Sept.-Oct. 1984:77–78.

Nelson, K., and D. Hedgecock. 1977. Electrophoretic evidence of multiple paternity in the lobster *Homarus americanus* (Milne-Edwards). *American Naturalist* 111:361–365.

Nielson, A. 1970. Trichoptera. In *Taxonomist's glossary of genitalia of insects*, ed. S. Tuxen. Darien, Conn.: S-H Service Agency.

Nielson, E. T. 1961. On the habits of the migratory butterfly *Ascia monuste* L. *Biologiske Meddelelser Kongelige Danske Videnskabernes Selskab* 23:1–81.

Noble, G. K. 1931. *The biology of the amphibia*. New York: McGraw-Hill.

Noble, G. K., and M. K. Brady. 1933. Observations on the life history of the marbled salamander, *Ambystoma opacum* Gravenhorst. *Zoologica* 11(8):89–132.

Norris, M. J. 1932. Contributions towards the study of insect fertility, I: the structure and operation of the reproductive organs of the genera *Ephestia* and *Plodia* (Lepidoptera, Phycitidae). *Proceedings of the Zoological Society of London* 3:595–611.

Nuttall, G. H., and G. Merriman. 1911. The process of copulation in *Ornithodorus moubata*. *Parasitology* 4(1):39–44.

Nuttall, G. H., and C. Warburton. 1911. *Ticks: a monograph of the Ixodidae*. Cambridge: Cambridge University Press.

Nyholm, T. 1969. Uber Bau und Funktion der Kopulationsorgane bei den *Cyphones* (Col. Helodidae) Studien uber die Familie Helodidae. X. *Entomologische Tidschrift Arg.* 90:233–271.

OConnor, B., and W. Reisen. 1978. *Chiroptoglyphus*, a new genus of mites associated with bats with comments on the family Rosensteiniidae (Acari: Astigmata). *International Journal of Acarology* 4(3):179–194.

O'Donald, P. 1980. Genetic models of sexual selection. Cambridge: Cambridge University Press.

Oldfield, G. N., I. M. Newell, and D. K. Reed. 1972. Insemination of protogynes of *Aculus cornutus* from spermatophores and description of the sperm cell. *Annals of the Entomological Society of America* 65:1080–1084.

Oldfield, G. N., R. F. Habza, and N. S. Wilson. 1970. Spermatophores of eriophyid mites. *Annals of the Entomological Society of America* 63:520–526.

Oliver, J. H. 1974. Reproduction in ticks. 3. copulation in *Dermacenter occidentalis* and *Haemaphysalis leporispalustris. Journal of Parasitology* 60:499–506.

Olson, E. C. 1981. The problem of missing links: today and yesterday. *Quarterly Review of Biology* 56:405–442.

Opell, B. D. 1979. Revision of the genera and tropical American species of the spider family Uloboridae. *Bulletin of the Museum of Comparative Zoology* 148:443–549.

———— 1983. The female genitalia of *Hyptiotes cavatus* (Araneae: Uloboridae). *Transactions of the American Microscopical Society* 102:97–104.

Ordnish, R. G. 1971. Entomology of the Aucklands and other islands south of New Zealand: Coleoptera: Hydraenidae. *Pacific Insects Monographs* 27:185–192.

———— 1974. Arthropods of the subantarctic islands of New Zealand: Coleoptera: Hydrophilidae. *Journal of the Royal Society of New Zealand* 4:307–314.

Organ, J. A., and L.A. Lowenthal. 1963. Comparative studies of macroscopic and microscopic features of spermatophores of some plethodontid salamanders. *Copeia* 1963(4):659–669.

Ossiannilsson, F., L. Russell, and H. Weber. 1970. Homoptera. In *Taxonomist's glossary of genitalia in insects,* ed. S. Tuxen. Darien, Conn.: S-H Service Agency.

Ostaff, D. P., J. H. Bordan, and R. F. Sheperd. 1974. Reproductive biology of *Lambdina fiscellaria lugubrosa* (Lepidoptera: Geometridae). *Canadian Entomologist* 106:659–665.

Otte, D. 1970. A comparative study of communicative behavior in grasshoppers. *Miscellaneous Publications of the Museum of Zoology* University of Michigan 141:1–168.

Outram, I. 1971. Aspects of mating in the spruce budworm, *Choristoneura fumiferana* (Lepidoptera: Tortricidae). *Canadian Entomologist* 103:1121–1128.

Ouye, M. T., R. S. Garcia, H. M. Graham, and D. F. Martin. 1965. Mating studies on the pink bollworm, *Pectinophora gossypiella* (Lepidoptera: Galechidae), based on the presence of spermatophores. *Annals of the Entomological Society of America* 58:880–882.

Page, R. E., and R. A. Metcalf. 1982. Multiple mating, sperm utilization, and social evolution. *American Naturalist* 119(2):263–281.

Pahnke, A. 1974. Zur biologie, okologie und anatomie einheimischer Halacaridae (Acari). Ph.D. thesis, University of Kiel.

Paperna, I. 1972. Monogenea of the Red Sea fishes, III: Dactylogyridae from littoral and reef fishes. *Journal of Helminthology* 46:47–62.

Parker, G. A. 1970. Sperm competition and its evolutionary consequences in insects. *Biological Reviews* 45:525–567.

———— 1979. Sexual selection and sexual conflict. In *Sexual selection and reproductive competition in insects,* eds. M. Blum and N. Blum. New York: Academic Press.

Passmore, N. I., and V. C. Carruthers. 1979. *South African frogs.* Johannesburg: Witwatersrand University.

Patton, J. L., and M. S. Hafner. In press. Biosystematics of the native rodents of the Galapagos archipelago. In *Patterns of evolution in Galapagos organisms,* ed. R. I. Bowman and R. E. Leviton. San Francisco: American Association for the Advancement of Science, Pacific Division.

Paulson, D. R. 1974. Reproductive isolation in damselflies. *Evolution* 23(1):40–49.

Pauly, F. 1952. Die "Copula" der Oribatiden (Moosmilben). *Naturwissenschaften* 39:572–573.

Pease, R. W. 1968. The evolutionary and biological significance of multiple pairing in Lepidoptera. *Journal of the Lepidopterist's Society* 22(4):197–209.

Peck, S. B. 1973. A systematic revision and the evolutionary biology of the *Ptomaphagus (Adelops)* beetles in North America (Coleoptera: Leiodidae: Catopinae) with emphasis on cave-inhabiting species. *Bulletin of the Museum of Comparative Zoology* 145:29–162.

———— 1981. Evolution of cave Cholevinae in North America (Coleoptera: Leiodidae). *Proceedings of the Eighth International Congress of Speleology,* ed. B. Beck. Bowling Green, Ohio.

———— 1983. Experimental hybridizations between populations of cavernicolous *Ptomaphagus* beetles (Coleoptera: Leiodidae: Cholevinae). *Canadian Entomologist* 115:445–452.

———— Manuscript. The distribution and evolution of cavernicolous *Ptomaphagus* beetles in the southeastern United States (Coleoptera: Leiodidae: Cholevinae) with new species and records.

Pennak, R. W. 1978. *The fresh-water invertebrates of the United States.* New York: Wiley and Sons.

Perez, R., and W. H. Long. 1964. Sex attractant and mating behavior in the sugarcane borer. *Journal of Economic Entomology* 57(4):688–690.

Peschke, K. 1979. Tactile orientation by mating males of the staphylinid beetle, *Aleochara curtula,* relative to the setal fields of the females. *Physiological Entomology* 4:155–159.

Picard, A. 1980. Spermatogenesis and sperm-spermatheca relations in *Spirorbis spirorbis* (L.). *International Journal of Invertebrate Reproduction* 2:73–83.

Piccioli, M. T. M., and L. Pardi. 1970. Studi sulla biologia de *Belanogaster* (Hymenoptera, Vespidae), I: Sull' Etogramma di *Belanogaster griseus* (Fab.). *Monitore Zoologico Italiano,* n.s., 3d supp. 9:197–225.

Picker, M. 1980. *Neoperla spio:* a species complex? *Systematic Entomology* 5:185–198.

Pickford, G. E. 1945. Le poulpe American: a study of the littoral octopoda of the western Atlantic. *Transactions of the Connecticut Academy of Arts and Sciences* 36:701–812.

Pine, R. H., D. C. Carter, and R. K. LaVal. 1971. Status of *Bauerus* van Gelder and its relationships to other nyctophiline bats. *Journal of Mammology* 52:663–669.

Pinto, J., and R. Selander. 1970. The bionomics of blister beetles of the genus *Meloe* and a classification of the New World species. *Illinois Biological Monographs* 42:1–222.

Pires, A. M. S. 1982. Taxonomic revision of *Bagatus* (Isopoda, Asellota) with a discussion of ontogenetic polymorphism in males. *Journal of Natural History* 16:227–259.

Platnick, N. I., and M. Shadab. 1976. On Colombian *Cryptocellus* (Arachnida, Ricinulei). *American Museum Novitates* 2605:1–8.

——— 1977. On Amazonian *Cryptocellus* (Arachnida, Ricinulei). *American Museum Novitates* 2633:1–17.

——— 1977. A new genus of the spider subfamily Gnaphosinae from the Virgin Islands (Araneae, Gnaphosidae). *Journal of Arachnology* 3:191–194.

——— 1978. A review of the spider genus *Anapis* (Araneae, Anapidae), with a dual cladistic analysis. *American Museum Novitates.* 2663:1–23.

——— 1979. A review of the spider genera *Anapisona* and *Pseudanapis* (Araneae, Anapidae). *American Museum Novitates* 2672:1–20.

——— 1980. A revision of the spider genus *Cesonia* (Araneae, Gnaphosidae). *Bulletin of the American Museum of Natural History* 165:335–386.

Platt, A. 1978. Editor's note. *Journal of the Lepidopterist's Society* 32(4):305.

Platt, A. P., S. D. Frearson, and P. N. Graves. 1970. Statistical comparisons of valval structure within and between populations of North American *Limenitis* (Nymphalidae). *Canadian Entomologist* 102(5):513–533.

Pliske, T. 1973. Factors determining mating frequencies in some New World butterflies and skippers. *Annals of the Entomological Society of America* 66(1):164–169.

Pope, C. H. 1941. Copulatory adjustment in snakes. *Field Museum of Natural History Zoological Series* 24:249–252.

Popp, R. 1967. Die Begattung bei den Vogelmilben *Pterodectes* Robin (Analgesoidea, Acari). *Zeitschrift für Morphologie Okologie Tiere* 59:1–32.

Popper, K. R. 1973. *Objective knowledge.* New York: Oxford University Press.

Post, D. C., and R. L. Jeanne. 1983. Relatedness and mate selection in *Polistes fuscatus* (Hymenoptera: Vespidae). *Animal Behaviour* 31:1261.

Prasad, M. 1974. Mannliche Geschlechtsorgane. *Handbuch der Zoologie,* 8 Band/51 Lieferung 9(2):1–15.

Price, R. D. 1966. The genus *Eomenopon* Harrison with descriptions of seven new species. *Pacific Insects* 8:17–28.

——— 1969. Two new species of *Eomenopon* Harrison (Mallophaga: Menoponidae) with a note on the structure of the genital sac. *Pacific Insects* 11:763–767.

——— 1971. A review of the genus *Holomenopon* (Mallophaga: Menoponidae) from the Anseriformes. *Annals of the Entomological Society of America* 64:633–646.

———— 1972a. Two new species of *Eomenopon* Harrison (Mallophaga: Menoponidae) from New Guinea lorikeets. *Pacific Insects* 14:23–26.

———— 1972b. Host records for *Geomydoecus* (Mallophaga: Trichodectidae) from the *Thomomys bottae-umbrinus* complex (Rodentia: Geomyidae). *Journal of Medical Entomology* 9:537–544.

———— 1974. Two new species of *Geomydoecus* from Costa Rican pocket gophers (Mallophaga: Trichodectidae). *Proceedings of the Entomological Society of Washington* 76:41–44.

———— 1975. The *Geomydoecus* (Mallophaga: Trichodectidae) of the southeastern USA pocket gophers (Rodentia: Geomyidae). *Proceedings of the Entomological Society of Washington* 77:61–65.

———— 1976. A new species of *Colpocephalum* (Phthiraptera) on *Threskiornis* (Aves) from Aldabra. *Systematic Entomology* 1:61–63.

Price, R. D., and J. R. Beer. 1964. Species of *Colpocephalum* (Mallophaga: Menoponidae) parasitic upon the Galliformes. *Annals of the Entomological Society of America* 57:391–402.

———— 1965. The *Colpocephalum* (Mallophaga: Menoponidae) of the Ciconiiformes. *Annals of the Entomological Society of America* 58:111–131.

Price, R. D., and T. Clay. 1972. A review of the genus *Austromenopon* (Mallophaga: Menoponidae) from the Procellariiformes. *Annals of the Entomological Society of America* 65:487–504.

Price, R. D., and K. C. Emerson. 1972. A new subgenus and three new species of *Geomydoecus* (Mallophaga: Trichodectidae) from *Thomomys* (Rodentia: Geomyidae). *Journal of Medical Entomology* 9:463–467.

———— 1982. A new species of *Colpocephalum* (Mallophaga: Menoponidae) from an Indian flamingo. *Journal of the Kansas Entomological Society* 47:63–65.

Price, R. D., and R. A. Hellenthal. 1975a. A review of the *Geomydoecus texanus* complex (Mallophaga: Trichodectidae) from *Geomys* and *Pappogeomys* (Rodentia: Geomyidae). *Journal of Medical Entomology* 12:401–408.

———— 1975b. A reconsideration of *Geomydoecus expansus* (Duges) (Mallophaga: Trichodectidae) from the yellow-faced pocket gopher (Rodentia: Geomyidae). *Journal of the Kansas Entomological Society* 48:33–42.

———— 1976. The *Geomydoecus* (Mallophaga: Trichodectidae) from the hispid pocket gopher (Rodentia: Geomyidae). *Journal of Medical Entomology* 12:695–700.

Pritchard, A. E., and E. W. Baker. 1955. *A revision of the spider mite family Tetranychidae.* Pacific Coast Entomological Society Memorial Series 2:1–472.

Prokopy, R. J., and G. L. Bush. 1973. Mating behavior of *Rhagoletis pomonella* (Diptera: Tephritidae), IV: courtship. *Canadian Entomologist* 105:873–891.

Prokopy, R. J., E. W. Bennett, and G. L. Bush. 1971. Mating behavior in *Rhagoletis pomonella* (Diptera: Tephritidae), I: site of assembly. *Canadian Entomologist* 103:1405–1409.

Proshold, F. I., L. E. La Chance, and R. D. Richard. 1975. Sperm production and transfer by *Heliothis virescens, H. subflexa,* and the sterile hybrid males. *Annals of the Entomological Society of America* 68(1):31–34.

Provost, M., and J. Haeger. 1967. Mating and pupal attendance in *Deinocerites cancer* and comparisons with *Opifex fuscus* (Diptera: Culicidae). *Annals of the Entomological Society of America* 60(3):565–574.

Purchon, R. D. 1977. *The biology of the Mollusca,* 2d ed. New York: Pergamon Press.

Putman, W. L. 1966. Insemination in *Balaustium* sp. (Erythraeidae). *Acarologia* 8(3):424–426.

Quick, H. E. 1947. *Arion ater* (L.) and *A. rufus* (L.) in Britain and their specific differences. *Journal of Conchology* 22(10):249–261.

———— 1960. British slugs (Pulmonata; Testacellidae, Arionidae, Limacidae). *Bulletin of the British Museum of Natural History* 6(3):105–226.

Raulston, J. R. 1975. Tobacco budworm: observations on the laboratory adaptation of a wild strain. *Annals of the Entomological Society of America* 68(1):139–142.

Raulston, J. R., J. W. Snow, H. M. Graham, and P. D. Lingren. 1975. Tobacco budworm: effect of prior mating and sperm content on the mating behavior of females. *Annals of the Entomological Society of America* 68(4):701–704.

Raven, R. J. 1984. A revision of the *Aname maculata* species group (Dipluridae, Araneae) with notes on biogeography. *Journal of Arachnology* 12:177–193.

Ray, C., and M. Williams. 1980. Description of the immature stages and adult male of *Pseudophilippia quaintaneii* (Homoptera: Coccoidea: Coccidae). *Annals of the Entomological Society of America* 73(4):437–447.

Reeve, M. R., and T. C. Cosper. 1974. Chaetognatha. In *Reproduction of marine invertebrates.* vol. 2, ed. A. Giese and J. Pearse. New York: Academic Press.

Reid, J. D. 1964. The reproduction of the sarcoglossan opisthobranch *Elysia maoria. Proceedings of the Zoological Society of London* 143:365–393.

Rentz, D. 1972. The lock and key as an isolating mechanism in katydids. *American Scientist* 60:750–755.

Restrepo-Mejia, R. 1980. Membracidos de Colombia -I. *Caldasia* 13(61):103–164.

Richards, A. M. 1970. Revision of the Rhaphidophoridae (Orthoptera) of New Zealand. Part XIII: A new genus from the Snares Islands. *Pacific Insects* 12:865–869.

———— 1971a. The Rhaphidophoridae (Orthoptera) of Australia. Part 9: The distribution and possible origins of Tasmanian Rhaphidophoridae, with description of two new species. *Pacific Insects* 13:575–587.

———— 1971b. The Rhaphidophoridae (Orthoptera) of Australia, Part 10: A new genus from southeastern Tasmania with New Zealand affinities. *Pacific Insects* 13:589–595.

———— 1974. Arthropoda of the subantarctic islands of New Zealand 7.

Orthoptera: Rhaphidophoridae. *New Zealand Journal of Zoology* 1:495–499.

Richards, O. W. 1927a. The specific characters of the British humblebees (Hymenoptera). *Transactions of the Royal Entomological Society of London* 75:233–265.

——— 1927b. Sexual selection and allied problems in the insects. *Biological Reviews* 2:298–360.

——— 1978. *The social wasps of the Americas.* London: British Museum (Natural History).

——— 1982. A revision of the genus *Belonogaster* de Saussure (Hymenoptera: Vespidae). *Bulletin of the British Museum of Natural History* (Entomology) 44(2):31–114.

Riddiford, L. M., and J. B. Ashenhurst. 1973. The switchover from virgin to mated behavior in female cecropia moths: the role of the bursa copulatrix. *Biological Bulletin* (Woods Hole) 144:162–171.

Ridley, M. 1978. Paternal care. *Animal Behaviour* 26:904–932.

Riemann, J., D. Mown, and B. Thorson. 1967. Female monogamy and its control in houseflies. *Journal of Insect Physiology* 13:407–418.

Rigby, J. 1963. Alimentary and reproductive systems of *Oxychilus cellarius* (Muller) (Stylommatophora). *Proceedings of the Zoological Society of London* 141:311–359.

Rindge, F. H. 1973. The Geometridae (Lepidoptera) of the Galapagos Islands. *American Museum Novitates* 2510:1–31.

Robertson, H. M., and H. E. H. Paterson. 1982. Mate recognition and mechanical isolation in *Enallagma* damselflies (Odonata: Coenagrionidae). *Evolution* 36:243–250.

Robinson, M. H. 1982. Courtship and mating behavior in spiders. *Annual Review of Entomology* 27:1–20.

Robinson, M. H., and B. Robinson. 1980. Comparative studies of the courtship and mating behavior of tropical araneid spiders. *Pacific Insect Monographs* 36:1–218.

Robinson, T. J. 1975. Contraception and sperm transport in domestic animals. In *The biology of spermatozoa,* ed. E. S. E. Hafez and C. G. Thibault. New York: S. Karger.

Roonwal, M. L. 1970. Isoptera. In *Taxonomist's glossary of genitalia of insects,* ed. S. H. Tuxen. Darien, Conn.: S-H Service Agency.

Rosen, D. E., and R. M. Bailey. 1963. The poeciliid fishes (Cyprinodontiformes), their structure, zoogeography, and systematics. *Bulletin of the American Museum of Natural History* 126:1–176.

Rosen, D. E., and M. Gordon. 1953. Functional anatomy and evolution of genitalia in peociliid fishes. *Zoologica* 38(1):1–47.

Rosenberg, H. I. 1967. Hemipenial morphology of some amphisbaenids (Amphisbaenia: Reptilia). *Copeia* 1967(2):349–361.

Roth, L. M. 1970. Evolution and taxonomic significance of reproduction in Blattaria. *Annual Review of Entomology* 15:75–96.

Roth, L. M., and B. Stay. 1961. Oocyte development in *Diploptera punctata* (Eschscholtz) (Blattaria). *Journal of Insect Physiology* 7:186–202.

Rothschild, M., and T. Clay. 1957. *Fleas, flukes and cuckoos*. New York: Macmillan.

Rowland, J. M., and J. R. Reddell. 1979. The order Schizomida (Arachnida) in the New World, I: Protoschizomidae and *dumitrescoae* group (Schizomidae, *Schizomus*). *Journal of Arachnology* 6:161–196.

Rowlands, I. W. 1958. Insemination by intraperitoneal injection. *Proceedings of the Society for the Study of Fertility* 10:150–157.

Rutowski, R. L. 1978. The courtship behaviour of the small sulphur butterfly, *Eurema lisa* (Lepidoptera: Pieridae). *Animal Behaviour* 26:892–903.

Ruttner, F. 1975. Ein Metatarsaler Halftapparat bei den Drohnen der Gattung *Apis* (Hymenoptera: Apidae). *Entomologica Germanica* 2(1):22–29.

Ryan, M. in press. *The tungara frog: a study of sexual selection and communication*. Chicago: University of Chicago Press.

Saint-Girons, H. 1975. Sperm survival and transport in the female genital tract of reptiles. In *The biology of spermatozoa*, ed. E. S. E. Hafez and C. G. Thibault. New York: S. Karger.

Sakaluk, S. K., and W. H. Cade. 1983. The adaptive significance of female multiple matings in house and field crickets. In *Orthopteran mating systems*, ed. D. Gwynne and G. Morris. Boulder, Colo.: Westview Press.

Samuelson, G. A. 1973. Alticinae of Oceania. *Pacific Insects Monographs* 30:1–165.

Santana, F. J. 1976. A review of the genus *Trouessartia*. *Journal of Medical Entomology* (suppl.) 1:1–128.

Sato, H. 1982. A new species of the genus *Dactylochelifer* (Pseudoscorpionidea: Cheliferidae) from Japan. *Acta Arachnologica* 30(2):105–110.

Schaller, R. 1968. *Soil animals*. Ann Arbor: University of Michigan Press.

———— 1971. Indirect sperm transfer by soil arthropods. *Annual Review of Entomology* 16:407–446.

Scherer, G. 1969. Die Alticinae des Indischen Subkontinentes. *Pacific Insects Monographs* 22:1–251.

Schroeder, P. C., and C. O. Hermans. 1975. Annelida: Polychaeta. In *Reproduction of marine invertebrates*, vol. 3, ed. A. Giese and J. Pearse. New York: Academic Press.

Schuster, I. J., and R. Schuster. 1970. Indirekte Spermubertragung bei Tydeidae (Acari, Trombidiformes). *Naturwissenschaft* 57:256.

Schuster, R., and I. J. Schuster. 1966. Uber das Fortpflanzungsverhalten von Anystiden-Mannchen (Acari, Trombidiformes). *Naturwissenschaft* 53:162.

———— 1969. Gestielte Spermatophoren bei Labidostomiden (Acari, Trombidiformes). *Naturwissenschaft* 56:145.

Scott, J. A. 1973a. Adult behavior and population biology of two skippers (Hesperiidae) mating in contrasting topographic sites. *Journal of Research in Lepidoptera* 12(4):181–196.

———— 1973b. Convergence of population biology and adult behavior in the sympatric butterflies, *Neominois ridingsii* (Papilionoida, Nymphalidae) and *Amblyscirtes simius* (Hesperiodidea, Hesperiidae). *Journal of Animal Ecology* 42:663–672.

—— 1974a. The interaction of behavior, population biology, and environment in *Hypaurotis crysalus* (Lepidoptera). *American Midland Naturalist* 91(2):383–394.

—— 1974b. Adult behavior and the population biology of *Poladras minuta*, and the relationship of the Texas and Colorado populations. *Pan-Pacific Entomologist* 50:9–22.

—— 1974c. Population biology and adult behavior of *Lycaena arota* (Lycaenidae). *Journal of the Lepidopterist's Society* 29(1):64–72.

—— 1978. Mid-valval flexion in the left valva of asymmetric genitalia of *Erynnis* (Hesperiidae). *Journal of the Lepidopterist's Society* 32(4):304–305.

Scott, J. A., and P. A. Opler. 1975. Population biology and adult behavior of *Lycaena xanthoides* (Lycaenidae). *Journal of the Lepidopterist's Society* 29(1):63–66.

Scudder, G. 1971. Comparative morphology of insect genitalia. *Annual Review of Entomology* 16:379–406.

Selander, R. 1964. Sexual behavior in blister beetles (Coleoptera: Meloidae), I: the genus *Pyrota*. *Canadian Entomologist* 96:1037–1082.

Selander, R. B., and J. M. Mathieu. 1969. Ecology, behavior, and adult anatomy of the *albide* group of the genus *Epicauta* (Coleoptera, Meloidae). *Illinois Biological Monographs* 41:1–168.

Severinghaus, L., B. Kurtak, and G. Eickwort. 1981. The reproductive behavior of *Anthidium manicatum* (Hymenoptera: Megachilidae) and the significance of size for territorial males. *Behavioral Ecology and Sociobiology* 9:51–58.

Shapiro, A. 1978. The assumption of adaptivity in genital morphology. *Journal of Research in Lepidoptera* 17(1):68–72.

Shear, W. A. 1972. Studies in the milliped order Chordeumida (Diplopoda): a revision of the family Cleidogonidae and a reclassification of the order Chordeumida in the New World. *Bulletin of the Museum of Comparative Zoology* 144(4):151–352.

—— 1976. The millipede family Conotylidae (Diplopoda, Chordeumida). Revision of the genus *Taiyutyla* with notes on recently proposed taxa. *American Museum Novitates* 2600:1–22.

—— 1981. The milliped family Tingupidae (Diplopoda, Chordeumatida, Brannerioidea). *American Museum Novitates* 2715:1–20.

Shelley, R. W. 1981. Revision of the milliped genus *Sigmoria* (Polydesmida: Xystodesmidae). *Memoirs of the American Entomological Society* 33:1–139.

Sherman, P. W. 1981. Electrophoresis and avian genealogical analysis. *Auk* 98(2):419–422.

Shields, O. 1967. Hilltopping. *Journal of Research in Lepidoptera* 6(2):69–178.

Shoop, C. R. 1960. The breeding habits of the mole salamander, *Ambystoma talpoideum* (Holbrook), in southeastern Louisiana. *Tulane Studies in Zoology* 8:65–82.

Short, R. V. 1979. Sexual selection and its component parts, somatic and genital selection, as illustrated by man and the great apes. *Advances in the Study of Behavior* 9:131–158.

Silberglied, R. E., J. Shepard and J. Dickinson. 1984. Eunuchs: the role of apyrene sperm in Lepidoptera? *American Naturalist* 123:255–265.

Silberglied, R. E., and O. R. Taylor. 1978. Ultraviolet reflection and its behavioral role in the courtship of the sulfur butterflies *Colias eurytheme* and *C. philodice* (Lepidoptera, Pieridae). *Behavioral Ecology and Sociobiology* 3:203–243.

Silvey, J. K. G. 1931. Observations on the life-history of *Rheumatobates rileyi* (Berg.) (Hemiptera-Gerridae). *Papers of the Michigan Academy of Science, Arts and Letters* 13:433–446.

Sivinski, J. 1980. Sexual selection and insect sperm. *Florida Entomologist* 63(1):99–111.

———— 1983. Predation and sperm competition in the evolution of coupling durations, particularly in the stick insect *Diapheromera veliei*. In *Orthopteran mating systems*, ed. D. Gwynne and G. Morris. Boulder, Colo.: Westview Press.

———— in press. Sperm in competition. In *Sperm competition and the evolution of animal mating systems*, ed. R. L. Smith. New York: Academic Press.

Smit, F. G. A. M. 1970. Siphonaptera. In *Taxonomist's glossary of genitalia of insects*, ed. S. Tuxen. Darien, Conn.: S-H Service Agency.

Smith, E. 1970. Evolutionary morphology of the external insect genitalia, II: Hymenoptera. *Annals of the Entomological Society of America* 63(1):1–27.

Smith, R. L. 1979. Repeated copulation and sperm precedence: paternity assurance for a male brooding water bug. *Science* 205:1029–1031.

———— 1980. Evolution of exclusive postcopulatory paternal care in the insects. *Florida Entomologist* 63(1):65–78.

Solignac, M. 1981. Isolating mechanisms and modalities of speciation in the *Jaera albifrons* species complex (Crustacea, Isopoda). *Systematic Zoology* 30:387–405.

Solomon, J. D., and W. W. Neel. 1973. Mating behavior in the carpenterworm, *Prionoxystus robiniae* (Lepidoptera: Cossidae). *Annals of the Entomological Society of America*. 66:312–314.

Southward, E. C. 1975. Pogonophora. In *Reproduction of marine invertebrates*, vol. 2, ed. A. Giese and J. Pearse. New York: Academic Press.

Speith, H . T. 1940. Studies on the biology of the Ephemeroptera, II: the nuptial flight. *Journal of the New York Entomological Society* 48:379–390.

———— 1968. Evolutionary implications of sexual behavior in *Drosophila*. In *Evolutionary biology*, ed. M. K. Hecht and W. C. Steere. New York: Appleton-Century-Crofts.

Spencer, K. A. 1969. The Agromyzidae of Canada and Alaska. *Memoirs of the Entomological Society of Canada* 64:1–308.

Spielman, A. 1966. The functional anatomy of the copulatory apparatus of male *Culex pipiens* (Diptera: Culicidae). *Annals of the Entomological Society of America* 59:309–314.

Spotila, J. R., and R. J. Blumer. 1970. The breeding habits of the ringed salamander, *Ambystoma annulatum* (Cope), in northwestern Arkansas. *American Midland Naturalist* 84(1):77–89.

Spratt, D. M. 1979. A taxonomic revision of the lungworms (Nematoda: Metastrongyloidea) from Australian marsupials. *Australian Journal of Zoology* (suppl. ser.) 67:1–45.

Spratt, D. M., and G. Varughese. 1975. A taxonomic revision of filarioid nematodes from Australian marsupials. *Australian Journal of Zoology* (suppl. ser.) 35:1–99.

Stein, D. S., and R. Hart. 1983. Brain damage and recovery: problems and perspectives. *Behavioral and Neural Biology* 37:185–222.

Stephenson, J. 1930. *The Oligochaeta.* Cambridge: Oxford University Press.

Stern, V. M., and R. F. Smith. 1960. Factors affecting egg production and oviposition in populations of *Colias philodice eurytheme* Boisduval (Lepidoptera: Pieridae). *Hilgardia* 29:411–454.

Sternlicht, M., and S. Goldenberg. 1971. Fertilization, sex ratio and postembryonic stages of the citrus bud mite *Aceria sheldoni* (Ewing) (Acarina, Eriophyidae). *Bulletin of Entomological Research* 60:391–397.

Sternlicht, M., and D. A. Griffiths. 1974. The emission and form of spermatophores and the fine structure of adult *Eriophyes sheldoni* Ewing (Acarina, Eriophyoidea). *Bulletin of Entomological Research* 63:561–565.

Sterrer, W. 1974. Gnathostomulida. In *Reproduction of marine invertebrates*, vol. 3, ed. A. Giese and J. Pearse. New York: Academic Press.

Sturm, H. 1956. Die Paarung beim Silberfischen *Lepisma saccharina*. *Zeitschrift für Tierpsychologie* 13(1):1–12.

Sublette, J., and W. W. Wirth. 1980. The Chironomidae and Ceratopogonidae (Diptera) of New Zealand's subantarctic islands. *New Zealand Journal of Zoology* 7:299–378.

Swailes, G. 1961. Laboratory studies on mating and oviposition of *Hylemya brassicae* (Bouche) (Diptera: Anthomyiidae). *Canadian Entomologist* 93:940–943.

Sweet, H. 1964. The biology and ecology of the Rhyparochrominae of New England (Heteroptera: Lygaeidae), II. *Entomologica Americana* 44:1–201.

Sweetman, H. L. 1938. Physical ecology of the firebrat, *Thermobia domestica* (Packard). *Ecological Monographs* 8(2):285–311.

Taberly, G. 1957. Observations sur les spermatophores et leur transfert chez les oribatés (Acariens). *Bulletin de la Societé Zoologique de France* 82:139–145.

Tauber, C. A. 1969. Taxonomy and biology of the lacewing genus *Meleoma*

(Neuroptera: Chrysopidae). *University of California Publications in Entomology* 58:1–93.

Taylor, O. R. 1967. Relationship of multiple mating to fertility in *Atteva punctella* (Lepidoptera: Yponomeutidae). *Annals of the Entomological Society of America* 60:583–590.

Tennessen, K. J. 1982. Review of reproductive isolating barriers in Odonata. *Advances in Odonatology* 1:251–265.

Thane, A. 1974. Rotifera. In *Reproduction of marine invertebrates*, ed. A. Giese and J. Pearse. New York: Academic Press.

Theodore, O. 1980. *Fauna Palestina Insecta, II — Diptera: Asilidae.* Jerusalem: Israel Academy of Sciences and Humanities.

Thornhill, R. 1976. Sexual selection and nuptial feeding behavior in *Bittacus apicalis* (Insecta: Mecoptera). *American Naturalist* 110:529–548.

———— 1980. Rape in *Panorpa* scorpionflies and a general rape hypothesis. *Animal Behaviour* 28(1):52–59.

———— 1983. Cryptic female choice and its implications in the scorpionfly *Harpobittacus nigriceps. American Naturalist:* 122:765–788.

Thornhill, R., and J. Alcock. 1983. *The evolution of insect mating systems.* Cambridge, Mass.: Harvard University Press.

Todd, E. L. 1972. Descriptive and synonymical notes for some species of Noctuidae from the Galapagos Islands (Lepidoptera). *Journal of the Washington Academy of Sciences* 62:36–40.

Tomlinson, W. E. 1966. Mating and reproductive history of blacklight-trapped cranberry fruitworm moths. *Journal of Economic Entomology* 59:849–851.

Toro, H., and E. de la Hoz. 1976. Factores mecánicos en la aislación reproductiva de *Apoidea* (Hymonoptera). *Revista de la Sociedad Entomológica de Argentina* 35:193–202.

Toschi, C. A. 1965. The taxonomy, life histories, and mating behavior of the green lacewings of Strawberry Canyon (Neuroptera: Chrysopidae). *Hilgardia* 36(11):391–431.

Townsend, J. I. 1971. Entomology of the Aucklands and other islands south of New Zealand: Coleoptera: Carabidae: Broscini. *Pacific Insect Monographs* 27:173–184.

Treat, A. E. 1975. *Mites of moths and butterflies.* Ithaca, N.Y.: Comstock Publishing Associates.

Trivers, R. L. 1972. Parental investment and sexual selection. In *Sexual selection and the descent of man, 1871–1971*, ed. B. Campbell. Chicago: Aldine.

Turner, J. R. G., C. A. Clarke, and P. M. Sheppard. 1961. Genetics of a difference in the male genitalia of east and west African stocks of *Papilio dardanus* (Lep.). *Nature* 191:935–936.

Tuxen, S. L., ed. 1970. *Taxonomist's glossary of genitalia of insects.* Darien, Conn.: S-H Service Agency.

Tyler, M. J. 1976. *Frogs.* London: Collins.

Ulagaraj, S. 1975. Mole crickets: ecology, behavior, and dispersion flight (Orthoptera: Gryllotalpidae: *Scapteriscus*). *Environmental Entomology* 4(2):265–273.

Uzzell, T. M. 1966. Teiid lizards of the genus *Neusticurus* (Reptilia, Sauria). *Bulletin of the American Museum of Natural History* 132(5):277–327.

———— 1969. Notes on spermatophore production by salamanders of the *Ambystoma jeffersonianum* complex. *Copeia* 1969:602–612.

Valerio, C. E. 1981. Spitting spiders (Araneae, Scytodidae, *Scytodes*) from Central America. *Bulletin of the American Museum of Natural History* 170(1):80–89.

van Demark, N. L., and R. L. Hays. 1952. Uterine motility responses to mating. *American Journal of Physiology* 170(3):518–521.

van der Vecht, J. 1971. The subgenera *Megapolistes* and *Stenopolistes* in the Solomon Islands. In *Entomological essays to commemorate the retirement of Professor K. Yasumatsu*. Tokyo: Hokuryukan Publishing Co.

van Helsdingen, P. J. 1965. Sexual behaviour of *Lepthyphantes leprosus* (Ohlert) (Araneida, Linyphiidae), with notes on the function of the genital organs. *Zoologische Mededelingen* 41:15–42.

van Tienhoven, A. 1968. *Reproductive physiology of vertebrates*. Philadelphia: W. B. Saunders.

Vardy, C. R. 1978. A revision of the neotropical wasp genus *Trigonopsis* Perty (Hymenoptera: Sphecidae). *Bulletin of the British Museum (Natural History) Entomology Series* 37:117–152.

Villavaso, E. 1975. Functions of the spermathecal muscle of the boll weevil, *Anthonomus grandis*. *Journal of Insect Physiology* 21:1275–1278.

Voss, R. 1979. Male accessory glands and the evolution of copulatory plugs in rodents. *Occasional Papers of the Museum of Zoology* University of Michigan 689:1–27.

Waage, J. K. 1979. Dual function of the damselfly penis: sperm removal and transfer. *Science* 203:916–918.

———— in press. Sperm competition and the evolution of odonate mating systems. In *Sperm competition and the evolution of animal mating systems*, ed. R. L. Smith. New York: Academic Press.

Walker, T. J. 1979. Calling crickets *(Anurogryllus arboreus)* over pitfalls: females, males and predators. *Environmental Entomology* 8(3):441–443.

Walker, W. 1980. Sperm utilization strategies in non-social insects. *American Naturalist* 115(6):780–799.

Wallace, M. M. H., and I. M. Mackerras. 1970. The entognathous hexapods. In *The insects of Australia,* ed. C.S.I.R.O. Melbourne: Melbourne University Press.

Wallace, M. M. H., and J. A. Mahon. 1972. The taxonomy and biology of Australian Bdellidae (Acari), I: subfamilies Bdellinae, Spinibdellinae and Cytinae. *Acarologia* 14:544–580.

———— 1976. The taxonomy and biology of Australian Bdellidae (Acari), II: subfamily Odontoscirinae. *Acarologia* 18(1):65–123.

Walton, W. 1960. Copulation and natural insemination. In *Marshall's Physiology of reproduction*, vol. 2 , ed. A. S. Parkes. London: Longman's.

Watanabe, M. 1960. *Cottidae*. Tokyo: Tokyo News Service.

Watson, J. A. L. 1966. Genital structure as an isolating mechanism in Odonata. *Proceedings of the Royal Entomological Society of London* (A) 41:171–4.

Watt, J. C. 1971. Entomology of the Aucklands and other islands south of New Zealand: Coleoptera: Scarabaeidae, Byrrhidae, Ptinidae, Tenebrionidae. *Pacific Insect Monographs* 27:193–224.

Webb, G. R. 1947. The mating-anatomy technique as applied to polygyrid landsnails. *American Naturalist* 81:134–147.

Webb, J. E. 1948. The evolution and host-relationships of the sucking lice of the Ferungulata. *Proceedings of the Zoological Society of London* 119:133–188.

Webber, H. H. 1977. Gasteropoda: Prosobranchia. In *Reproduction of marine invertebrates*, vol. 4, ed. A. Giese and J. Pearse. New York: Academic Press.

Weems, H. V. 1981. Foreword. *Occasional Papers of the Florida State Collection of Arthropods* 2:iii–iv.

Wells, M. J. 1978. *Octopus*. London: Chapman and Hall.

Wells, M. J., and I. Wells. 1977. Cephalopoda: Octopoda. In *Reproduction of marine invertebrates*, vol. 4, ed. A. Giese and J. Pearse. New York: Academic Press.

Werneck, F. L. 1950. *Os Malofagos de mamiferos, II: Ischnocera (continuacão de Trichodectidae) y Rhyncophthirinae*. Rio de Janeiro: Instituto Oswaldo Cruz.

West-Eberhard, M. J. 1969. The social biology of polistine wasps. *Miscellaneous Publications of the Museum of Zoology* University of Michigan 140:1–101.

——— 1983. Sexual selection, social competition, and speciation. *Quarterly Review of Biology* 58:155–183.

——— 1984. Sexual selection, social communication, and species specific signals in insects. In *Insect communication*, ed. T. Lewis. London: Academic Press.

Westheide, W. 1967. Monographie der Gattung *Hesionidas* Friedrich und *Microphthalmus* Mecznikow (Polychaeta, Hesionidae). *Zeitschrift für Morphologie Oekologie Tiere* 61:1–159.

Weygoldt, P. 1969. *The biology of pseudoscorpions*. Cambridge, Mass.: Harvard University Press.

——— 1970. Vergleichende Untersuchungen zur Fortpflanzungsbiologie der Pseudoscorpione, II. *Zeitschrift für Zoologische Systematik Evolutionsforschung* 8:24–259.

——— 1972. Spermatophorenbau und Samenubertragung bei Uropygen (*Mastigoproctus brasilianus* C. L. Koch) und Amblypygen (*Charinus brasilianus* Weygoldt und *Admetus pumilio* C. L. Koch) (Chelicerata, Arachnida). *Zeitschrift für Morphologie Oekologie Tiere* 71:23–51.

———— 1977. Communication in crustaceans and arachnids. In *How animals communicate,* ed. T. A. Sebeok. Bloomington: Indiana University Press.

Weygoldt, P., A. Weisemann, and K. Weisemann. 1972. Morphologische-Histologische Untersuchungen an den Geschlechtsorganen der Amblypygi unter besonderer Berücksichtigung von *Tarantula marginemaculata* C. L. Koch (Arachnida). *Zeitschrift für Morphologie der Tiere* 73:209–247.

White, M. J. D. 1978. *Modes of speciation.* San Francisco: W. H. Freeman.

White, R. E. 1968. A review of the genus *Cryptocephalus* in America north of Mexico (Chrysomelidae: Coleoptera). *U.S. National Museum Bulletin* 290:1–124.

Wickler, W. 1966. Ursprung und biologische Deutung des Genitalprasentierens mannlicher Primaten. *Zeitschrift für Tierpsychologie* 23:422–437.

———— 1969. Socio-sexual signals and their intraspecific imitation among primates. In *Primate ethiology,* ed. D. Morris, pp. 89–189. New York: Doubleday.

Wigglesworth, V. B. 1965. *The principles of insect physiology.* London: Methuen.

Wiley, M. L., and B. B. Collette. 1970. Breeding tubercles and contact organs in fishes: their occurrence, structure, and significance. *Bulletin of the American Museum of Natural History* 153:143–216.

Williams, F., and J. Campbell. 1980. *Echinobothrium bonasum,* a new species of cestode from the cownose ray. *Journal of Parasitology* 66:1036–1038.

Williams, G. E. 1966. *Natural selection and adaptation.* Princeton: Princeton University Press.

Williams, J. L. 1941. The relations of the spermatophore to the female reproductive ducts in Lepidoptera. *Entomological News* 52(3):62–65.

Wilson, E. O. 1971. *The insect societies.* Cambridge, Mass.: Harvard University Press.

Wing, S. R. 1982. The reproductive ecologies of three species of fireflies. Master's thesis, University of Florida, Gainesville.

Wirth, W. W. 1969. New species and records of Galapagos Diptera. *Proceedings of the California Academy of Sciences,* 4th ser. 36:571–594.

Wirth, W. W., and J. E. Sublette. 1970. A review of the Podonominae of North America with descriptions of three new species of *Trichotanypus* (Diptera: Chironomidae). *Journal of the Kansas Entomological Society* 43:335–354.

Witte, H. 1975. Funktionsanatomie der Genitalorgane und Fortpflanzungsverhalten bei den Mannchen der Erythraeidae (Acari, Trombidiformes). *Zeitschrift für Morphologie Tiere* 80:137–180.

———— 1977. Bau der Spermatophore und funktionelle Morphologie der männlichen Genitalorgane von *Sphaerolophus cardinalis* (C. L. Koch) (Acarina, Prostigmata). *Acarologia* 19:74–81.

Wocjik, D. P. 1969. Mating behavior of 8 stored-product beetles (Coleoptera: Dermestidae, Tenebrionidae, Cucujidae, and Curculionidae). *Florida Entomologist* 52(3):171–197.

Woodring, J. P. 1969. Observations on the biology of six species of acarid mites. *Annals of the Entomological Society of America* 62(1):102–108.

Woodring, J. P., and E. F. Cook. 1962. The biology of *Ceratozetes cisalpinus* Berlese, *Scheloribates laevigatus* Koch, and *Oppia neerlandica* Oudemans (Oribatei), with a description of all stages. *Acarologia* 4(1):101–137.

Wooldridge, D. P. 1969. New species of *Paracymus* from Mexico and Central America. *Journal of the Kansas Entomological Society* 42(4):413–421.

Woyke, J. 1964. Causes of repeated mating flights by queen honeybees. *Journal of Agricultural Research* 3(1):17–23.

Wright, K. A. 1978. Structure and function of the male copulatory apparatus of the nematodes *Capillaria hepatica* and *Trichuris muris*. *Canadian Journal of Zoology* 56:651–662.

Yamaguti, S. 1961. *Systema Helminthum*, vol. 3, pt. I. New York: Interscience.

Yanagimachi, R., and M. C. Chang. 1963. Sperm ascent through the oviduct of the hamster and rabbit in relation to the time of ovulation. *Journal of Reproduction and Fertility* 6:413–420.

Young, F. 1981. Predaceous water beetles of the genus *Desmopachria:* the *convexagrana* group (Coleoptera: Dytiscidae). *Occasional Papers of the Florida State Collection of Anthropods* 2:1–11.

Young, J. H. 1968. The morphology of *Haemogamasus ambulans*, II: reproductive system (Acarina: Haemogamasidae). *Journal of the Kansas Entomological Society* 41:532–543.

Zouros, E., and C. B. Krimbas. 1970. Frequency of female digamy in a natural population of the olive fruit fly *Dacus oleae* as found by using enzyme polymorphism. *Entomologia experimentalis et applicata* 13:1–9.

Zug, G. R. 1966. The penial morphology and relationships of cryptodiran turtles. *Occasional Papers of the Museum of Zoology* University of Michigan 647:1–24.

Zwick, P. 1982. A revision of the oriental stonefly genus *Phanoperla* (Plecoptera: Perlidae). *Systematic Entomology* 7:87–126.

Zwolfer, H. 1974. Das Treffpunkt-Prinzip als Kommunikationsstrategie und isolationsmechanismus bei Bohrfliegen (Diptera: Trypetidae). *Entomologica Germanica* 1:11–20.

Index